New Frontiers in Regional Science: Asian Perspectives

Volume 15

New Frontiers in Regional Science: Asian Perspectives

This series is a constellation of works by scholars in the field of regional science and in related disciplines specifically focusing on dynamism in Asia.

Asia is the most dynamic part of the world. Japan, Korea, Taiwan, and Singapore experienced rapid and miracle economic growth in the 1970s. Malaysia, Indonesia, and Thailand followed in the 1980s. China, India, and Vietnam are now rising countries in Asia and are even leading the world economy. Due to their rapid economic development and growth, Asian countries continue to face a variety of urgent issues including regional and institutional unbalanced growth, environmental problems, poverty amidst prosperity, an ageing society, the collapse of the bubble economy, and deflation, among others.

Asian countries are diversified as they have their own cultural, historical, and geographical as well as political conditions. Due to this fact, scholars specializing in regional science as an inter- and multi-discipline have taken leading roles in providing mitigating policy proposals based on robust interdisciplinary analysis of multifaceted regional issues and subjects in Asia. This series not only will present unique research results from Asia that are unfamiliar in other parts of the world because of language barriers, but also will publish advanced research results from those regions that have focused on regional and urban issues in Asia from different perspectives.

The series aims to expand the frontiers of regional science through diffusion of intrinsically developed and advanced modern regional science methodologies in Asia and other areas of the world. Readers will be inspired to realize that regional and urban issues in the world are so vast that their established methodologies still have space for development and refinement, and to understand the importance of the interdisciplinary and multidisciplinary approach that is inherent in regional science for analyzing and resolving urgent regional and urban issues in Asia.

Topics under consideration in this series include the theory of social cost and benefit analysis and criteria of public investments, socio-economic vulnerability against disasters, food security and policy, agro-food systems in China, industrial clustering in Asia, comprehensive management of water environment and resources in a river basin, the international trade bloc and food security, migration and labor market in Asia, land policy and local property tax, Information and Communication Technology planning, consumer "shop-around" movements, and regeneration of downtowns, among others.

More information about this series at http://www.springer.com/series/13039

Zheng Wang

Principles of Regional Science

Springer

Zheng Wang
Chinese Academy of Sciences
Institutes of Science and Development
Beijing, China

ISSN 2199-5974 ISSN 2199-5982 (electronic)
ISBN 978-981-10-5366-5 ISBN 978-981-10-5367-2 (eBook)
DOI 10.1007/978-981-10-5367-2

Library of Congress Control Number: 2017953014

Printed on acid-free paper

This Springer imprint is published by Springer Nature
The registered company is Springer Nature Singapore Pte Ltd.
The registered company address is: 152 Beach Road, #21-01/04 Gateway East, Singapore 189721, Singapore

Preface

In 1954, when Walter Isard and other scholars proposed to build a regional science, the "region" had been made into a scientific geography concept, had infiltrated the economy and geography, and had begun to develop. Today, regional science forms an independent science discipline. With the help of economic principles and methods, regional science has adopted the analytical methods of modern economics and the spatial analysis mode of modern geography. Many scholars from geography became regional scientists. In the early 1980s, influenced by the development of regional science, these scholars clearly proposed that two geographical images appear: regional and environment. This understanding makes the scientific concept of the region independent from that of geography. In this way began the carrier–place development of a rational concept, as the concept of the particle is the object of the abstract of mass–object, which formed particle dynamics. "Region" also supports a new discipline, as an abstraction of local economics, a spatial link, and raises new issues such as regional governance and regional cooperation. For such changes, geographers, economists, and management scientists have had to work hard.

In the early 1950s, after China accepted the Soviet school of geography, regional science had not been popularized and had not been applied. However, because China has such a vast geographical area, regional research has become a long-term research topic for geographers and economists. The economic geographer Chuanjun Wu emphasizes "territory" research, in fact, from the structural point of view of the nature of the region, and the female geographer Zhongmin Yan, in her study of urban systems, is the pioneer of international regional science in China. Wuyang Yang has come from economics, with the perspective of theoretical geography, to study location theory, which is China's theoretical regional research banner. All these scholars have facilitated the spread of regional science in China and contributed to regional science in accordance with the reality of China.

The author of this book was originally a Ph.D. in geomorphology. In the late 1980s, inspired by logical regional scientific thinking, in the face of China's regional development needs, the author entered regional scientific research. In this work, the author in regional science must begin a new phase of learning, studying, and understanding. Combined with my own work, I wrote the book *Principles of Regional Science* in 1994. The books had my initial sense as a Chinese scholar in the field of regional science. Today, I have modified this approach for more regional scientists. In this volume, Dr. Changxin Liu and Dr. Yi Sun completed the writing about the CGE method in regional science, Chap. 7, which enriches this book; Dr. Changxin Liu also participated in the revision of the old manuscript; and my graduate student Ms. Hailing Zhang was involved in translating the original Chinese version of this book into the English version. The author thanks them all for their work. In a sense, they should also be the authors of this book.

At the time of publication of this book, I wish to thank Professor Kingsley Haynes of GMU (George Mason University), Professor J.C. Thill of the University of North Carolina, and Professor Higano of the University of Tsukuba for their long-term cooperation and help. Without them, the author could not have completed this book.

Beijing, China Zheng Wang

Contents

Chapter 1
Introduction

Abstract This first chapter is about the guidance of regional science; In this chapter, the author expounds his own understanding of the concept of the region, emphasizing that the region resembles concepts of physics, such as the concept of a regional, local abstraction of these geographical entities. The author also discusses the concept of space from the perspective of regional analysis of the academic significance of space. The author argues that these discussions are the philosophical basis for the elaboration of regional scientific principles.

Space and Region are the core concept of geography. Space refers to the division of the Earth's surface, which is essentially a mathematical abstraction of the Earth's atmosphere. Area is the specialization of space, this specialization is due to the area of population, resources, environment and other geographical factors occurred. As a scientific concept, the area is abstracted as "region". So we need to discuss the concept of "space".

Typically, in geography, especially economic geography, space is understood to be two dimensional. On one hand, the concept of space in geography is mainly used in the fields of economic geography and general human geography. On the other hand, when referring to the concept of space in geography, space generally has the sense of both economic geography and general human geography. Although the concept of space is also applied in other disciplines, such as geomorphology and climatology, it has, more or less, some physical significance. In this sense, the latter is not a two-dimensional mathematical abstraction on the surface of the Earth. In this book, *space* particularly refers to a two-dimensional substrate in economic geography and in the general human geography sense of the spatial process, especially the regional (science) process, as also in the fields of economic geography and general human geography.

In the traditional study aspect of the spatial geography process, as it involves economic processes, the idea that "The first is location, the second is location, the third is still location" was popular in the 1930s.Thus, the common analysis mode in the space process model is Operations Research. However, the "location first" thinking has been disrupted. In modern geography, the motion and diffusion of population and capital in the Earth's space are often used as the basic dynamic factors of spatial processes. Because of the discontinuities and actors of geographical space, a general analytical model of continuous space is difficult to describe for

Z. Wang, *Principles of Regional Science*, New Frontiers in Regional Science: Asian Perspectives 15, DOI 10.1007/978-981-10-5367-2_1

the geospatial process. Therefore, idea and models based on statistical mechanics are also mainstream models in the analysis of geography spatial processes.

The concept of spatial process was introduced by Friedrich Ratzal. After years of development, the spatial process of modern geography refers to evolution and development of spatial contact, spatial pattern, and spatial form, which is the space kinetics of economic geography phenomena. In essence, space filled with regions is an abstract of the Earth's space, so space is always linked with the regions. However, it is necessary to strictly distinguish the spatial economic process and the regional economic process, which, respectively, emphasize the effect of spatial location on an economic process and the economic process of regions hosting the economy. Usually, spatial economic process and the regional economic process are collectively called the economic geography process. We can understand that the former describes the microeconomic phenomenon, in which the enterprises are the main target, and the latter describes the macroeconomic phenomenon, in which the regions having macroeconomic characteristics are the main target. The term 'local' indicates a mesoscape feature.

The concept of space in geography derives formed a series of concepts. (1) Space form refers to the shape of space occupied or the spatial distribution characteristics of geographical objects. (2) Spatial distribution refers to the points and their statistical relationships on Earth's surface or in its abstract form, the space of geographical objects. (3) Spatial relationships often refer to the interlinkages of distribution units. (4) Spatial structure usually, from the perspective of the landscape, is a division of space and an abstract of the relationship among points, lines, and surfaces. These concepts are still preserved in regional sciences, but the use of the term geographical structure in the Chinese regional scientific literature encompasses the dual meaning that the structure of the region as an economy and the integration of the region as a natural environment the geographical structure is often cited, although the term is often equated with the West's regional spatial structure. However, because of the importance of Chinese geography to the physical geography structure, a geographical structure is used to refer to land structure and the topography unit structure, which is mainly used in economic geography. Sometimes, geographers also adopt the spatial configuration to describe special types of general spatial structure. This term is used to refer to the route in the form of spatial structure units and also means some kinds of geographical division of pure forms.

The movement and diffusion of a population and capital cause the phenomena of migration, aggregation, anti-aggregation (disaggregation, also for divergence), growth, and clustering of population and capital. With the recent development of the Geographic Information System (GIS), spatial analysis breaks through the limitations of economy and population issues, which involve some natural geographical phenomena, but in this book, space is strictly limited to the economic geography process. In this process, population and enterprise movement having capital characteristics are the basic movements. Location choice is the most typical need of microeconomic geography.

With the development of the new economic geography of Krugman, the studies of geography spatial process theory and spatial economics theory appear convergent. In fact, the monograph of spatial economics by Fujita et al. (1999) obviously contributes to theoretical geography.

The space (economic) process is concerned with the economic behavior of enterprises in space. However, in geography, the synthesis of a large number of enterprises is another economic body, the region, which is a macro-economy of geography. As a matter of geography, enterprise and region are different. Enterprises can be mobile and choose their own optimal position, which forms the industrial location. In contrast to the enterprise, the location of the region cannot be mobile. It does not concern profit but rather macroeconomic issues such as growth, employment, finance, and trade. Enterprise pursues profits when it selects the location and gathers in space. Therefore, enterprises corresponding to the spatial problem raise a microeconomic geography question. Regional concerns of growth and development are macroeconomic issues. Differing from the general macroeconomic process, the regional process also concerns regional competition and cooperation, and regional development of populations, resources, environment coordination issues, and geopolitical relationships, which constitute the macroeconomic geographical features of the region. In the traditional theory, spatial economics and regional economics are not clearly distinguished, but they are facing two different economies, enterprise and region, respectively, which lead to different research starting points. This understanding also constitutes the epistemological basis of distinguishing microeconomic geography and macroeconomic geography. On the other hand, the study of regional processes in economics have formed relatively independent disciplines after absorbing some of the ideas and methods of geography, namely, regional science (Isard 1975), and the traditional geography theory and method then constitutes regional economic geography (Chen Cai et al. 1991). At present, the two disciplines are more and more convergent. One example is that the National Science Foundation of the United States combines geography and regional science, which is stated as geography and regional science.

The regional process is the manifestation of macroeconomic geography. Its most important features are growth behavior and the behavior of people and the land. The first visual angle of the analysis of the regional process is growth behavior under the constraint of the human–land relationship. The "land" of man and land is characterized by the structure of the region, which includes the resource environment structure, population structure, urban and rural structure, spatial structure, regional economic system structure, and geographical structure. In research, this is the convergence of economic growth research methods. The second perspective is the coexistence of multiple regions. The study of geography is multi-regional economics. This method needs to break with the classic macroeconomic research methods. Actually, the development of multi-regional economics, the major economical development in the late twentieth century, has provided exposure to the main community in the form of macroeconomic geography. The modeling methods of these two perspectives are distinctly focused on the application of the stability theory of differential equations.

Chapter 2
Location Theory

Abstract The author argues that location theory is the basic theory of regional science, resembling the status of statistics in physics, which expounds the basic concept of regional science and abstracts rational analysis from local and geographical analyses. However, to clarify the location theory, more basic content, interregional interactions, that is, a spatial interaction, also need to be clarified. Thus, in this chapter the author presents a discussion of spatial interactions. One of the main developments is to use statistical physics to emphasize that the empirical spatial interaction is an exponential form of attenuation, which is a development of Wilson's theory of interaction. In this chapter, the author discusses the study of location, trying to give an analytical interpretation of location theory.

2.1 Spatial Interaction

2.1.1 Newton Potential

Space exists in a number of places or cities and has certain resources and populations. In the case of space access, capital and population have contracted because of the presence of the market, employment opportunities, etc. This link is expressed as the interaction among areas, which is the space interaction. The measurement of spatial interaction is generally taken as a commodity flow, capital flow, and population flow. When the population (products) of place A reach place B, we can measure the effect of the numbers of A to B. Here there have been two issues. One is that the effect of A on B is not necessarily equal to the effect of B on A. The population (products flows) intensity from place A to place B is considered as a value, and population B (products flows) does not necessarily reach A with the same number. "Force" does not mean "reaction." The other issue is that the total population of A and B in two places is dynamic, and even a one-way flow. The interaction is the economic relationship between demand and supply. Once again, this interaction is not stable. It is not instantaneous, but rather the result of statistical averaging; this average value has to be calculated. It is necessary to accept the statistical physics of the state after the assumption, which is contrary to our understanding of the interaction force in physics. To overcome this difficulty, spatial interaction does not require a similar "Newton's third law" be established.

Z. Wang, *Principles of Regional Science*, New Frontiers in Regional Science: Asian Perspectives 15, DOI 10.1007/978-981-10-5367-2_2

The frequency or probability of flow can measure the interaction to eliminate the influence of spatial system evolution and fluctuations. The reference frequency is replaced by the flow, which means that the "force" of the interaction of the geography space is the result of the statistics. Smith (1987) introduced an axiomatic system to try to solve the problem of the rationality of statistical measure.

In long-term development, it has been found that spatial interaction has the form of distance decay. Spatial interaction has the characteristics of the attenuation of the distance, which is the first law of geography for some scholars. So, the Newton gravity formula is suggested to use to measure interactions:

$$F = -G\frac{mM}{r^2} \tag{2.1.1}$$

Here m is the demand for B and M is the supply of A. G is an empirical constant. When m, M are interpreted as two of the populations, G is the interaction coefficient, having a role of normalization in a system with many local connections. Here m and M are explaining the demand and supply, which is important, because they show the nonsymmetry of space interaction.

In numerical calculation the use of formula (2.1.1) is not reasonable. The following form was developed later (by Haynes and Fotheringham 1984):

$$F = -G\frac{m^{a_1}M^{a_2}}{r^b} \tag{2.1.2}$$

This model can modify the parameters according to the actual conditions, but the increase of the parameters contains the analysis of the problem, because the change of B exerts a large interference to F and how to state the parameters becomes difficult.

Applying the (2.1.1) type can be roughly discussed as two local A and C divisions of market space. Set the demand to be uniformly filled with the entire space: A, C distance is D. Then, the space is bound to have a point r_0 (a point to zero) so that it can make the next equation setup:

$$G\frac{M_a}{r_0^2} = G\frac{M_c}{(d - r_0)^2} \tag{2.1.3}$$

This point can be considered the equilibrium point of the interaction of space. Simply reformate it as

$$r_0 = d/\left(1 + \sqrt{M_c/M_a}\right) \tag{2.1.4}$$

The significance of this result is that the spatial interaction can produce a "potential" in the space, which means the location and spatial structure, namely, "space competition." The type (2.1.4) is determined by the equilibrium point position, which sometimes is referred to as the "break point." Haynes and Fotheringham (1984) give some enlightening examples.

2.1.2 Maximum Entropy Model

The Newton model of space interaction experience, the potential theory (Wang 1983), proved that the two-dimensional space cannot exist in the Newton potential (2.1.1) style, which encountered fundamental difficulties. In fact, Wilson (1967) derived the first theoretical model from the maximum entropy principle: the work of Wilson has brought fundamental progress to the study of space interaction.

Wilson set that several places exist in the space. O_j represents total supply of place j. D_k represents the demands of place k. T_{ik} represents the space flow of place j to place k. C_{jk} represents the current cost of place j to place k, namely, a generalized distance. C is total cost of flow. The flow in this area has the following constraints:

$$\sum_k T_{jk} = O_j, j = 1, \ldots m \tag{2.1.5}$$

$$\sum_j T_{jk} = D_k, k = 1, \ldots n \tag{2.1.6}$$

$$\sum_j \sum_k T_{jk} C_{jk} = C, j = \overline{1, m}, k = \overline{1, n} \tag{2.1.7}$$

Wilson thought that the entropy of the system should tend to be the largest in the condition of the closed system. At this time, the system can achieve a stable state, and the entropy of the system is

$$S = \ln \Omega \tag{2.1.8}$$

Within the system, Ω is the distribution function, defined as the multiplication symbol $T!/\Pi_{j,k} T_{jk}, \Pi$. $T = \sum T_{jk}, \Omega$ represents the system's microscopic state. The regional system is the regional system for a statistical physics of the thermodynamic system, a or macro-system. Wilson considered (2.1.8) the state distribution of the maximum system state distribution as the most likely state. The proof of the process is described by Wilson (1967), and a brief proof is also provided by (Wang et al. 2002). As a statistical characteristic, it can be considered as $\langle T_{jk} \rangle$. In (2.1.5), (2.1.6), (2.1.7), and (2.1.8), the maximum value of the state is obtained by using the Lagrange method.

$$\langle T_{jk} \rangle = \exp(-\lambda_j) \exp(-v_k) \exp(-\beta C_{jk}) \tag{2.1.9}$$

$\langle T_{jk} \rangle$ is T_{jk}. $\exp(-\lambda_j) \exp(-v_k)$ is constant and T_{jk} will be normalized.

$$T_{jk} = A_j B_k O_j D_k \exp(-\beta C_{jk}) \tag{2.1.10}$$

$$A_j = \left[\sum_k B_k D_k \exp(-\beta C_{jk})\right]^{-1} \tag{2.1.11}$$

$$B_k = \left[\sum_j A_j O_j \exp(-\beta C_{jk})\right]^{-1} \tag{2.1.12}$$

Equation (2.1.10) represents the spatial interactionT_{jk}. Two characteristics exist. First, it is the result of statistics; the numbers n and m are required to be very large, so it is the macro-representation of geography, although in the field of economics it belongs to microeconomics. Second, the possibility of the isolated system to achieve the maximum entropy is meaningful, only considering the commodity exchange, for the system is closed to the maximum entropy. From a global perspective, the commodity system is closed. Therefore, the space interaction can have the same decay form of $\exp(-\beta C_{jk})$. The coefficient A_j , B_k is indeed related to the regional system. In future, we consider $f(C_{jk})$ as the nuclear interaction in space. We know that the problem is complicated. We cannot think that the world has a unified value or measurement level, so Wilson's theory can only be been called the spatial interaction between cities or places in "a world of global closed sense."

In practical work, we often consider some simplified forms. In addition, provide demand or excess demand, supply shortage, and insufficient circulation. Then, give up constraint (2.1.6). There can be

$$T_{jk} = D_k A_j O_j f(C_{jk}) \tag{2.1.13}$$

Here $A_j^{-1} = \sum_k D_k f(C_{jk})$. Provide sufficient or excess supply, insufficient demand, and limited circulation costs. There can be

$$T_{jk} = O_j B_k D_k f(C_{jk}) \tag{2.1.14}$$

In (4), $B_k^{-1} \sum_j O_j f(C_{jk})$. Under the condition of no constraint, the space interaction is

$$T_{jk} = P_j P_k f(C_{jk}) \tag{2.1.15}$$

P_i is a standard factor and proportional to supply O_j or demand D_k. Here, P_i is the population of place j. It implies the supply O_j is the proportional function of the population P_i, and the demand $D_k$is the proportional function of the population P_i. This is a special kind of production function and demand function.

In later development, Wilson (1974) considered the imbalance between supply and demand and modified the interaction of the spatial relationships:

$$T_{jk} = A_j B_k O_j g(D_k) f(C_{jk}) \tag{2.1.16}$$

Here $g(D_k) = D_k^{\alpha}, \alpha > 0$ and α is a normal number. In particular, Wilson introduced the probability of interaction; this opened up a new way for the development of theory.

2.1.3 Agent-Based Model (Population Model)

Wang Zheng et al. (1986) and Wang explained the spatial interaction as the population, capital, etc., in the macro-performance of the Brownian movement in a two-dimensional space. Wang thought that the population and other particles were lively. They called this kind of life particles as population particles. The mouth of the particle is actually a kind of agent, so the mouth of the particle model could be called a model based on the principle, or it is agent based. Because the body is a physical entity, different from the general physical particles, as the body can have a unique personality, there is autonomy and it thus may be identified. From this point of view, the general physical particle is a special subject, which can be identified when its body has disappeared. The general agent-based model is characterized by a large number of individual components and the overall performance of the statistical characteristics. This behavior from the microscopic behavior of the macro-behavior of the process is known as "emerging." The particles of the autonomous body have the characteristics of life, but the diffusion of the body is subject to Brownian movement. In considering the life and migration of the population, introduce the mobility lifetime distribution:

$$g(t) = t^{\frac{1}{2}} \exp(-at) \tag{2.1.17}$$

So Wang Zheng derived that the space interaction has a basic form, whose derivation was given in the appendix of this chapter.

$$f(r) = k\exp(-\beta r) \tag{2.1.18}$$

Here $f(r)$ is the interaction nucleus. r is the generalized distance measure. β was found to be $\sqrt{4a\pi T/h^2}$. The h^2 is the area of the smallest unit of space and is considered as the domain element. T is time that spreads from one domain to another or the residence time in the domain. The size of the domain is related to the observation of the geographical processes. For population diffusion, the diffusion of the countryside is a village in the area. The urban area is spread by the district. So, the interaction of various diffusion leads to a value that is not consistent. Therefore, the spatial interaction of different regions is not the same. But for a certain geographical process, domain elements are identified. From $\beta \approx \sqrt{T/h^2}$, when h^2 becomes larger, the interregional interaction can be enhanced. The entrance may flow in two places, but not in the two villages, because the domain of the city is much larger than that of the country. For tourism geography, according to the actual data for Li Shan and Wang Zheng (2010) estimates that the provincial scale parameter value of the space is 0.00065. In the field of tourism, the spatial scale interaction damping value is 0.00236. By the same method, we can obtain the damping parameter value of the county area element space at 0.00527. From the interaction of tourism space, the smaller the damping value, the slower the attenuation, which shows that "strange effect" of tourist destination: the farther from the source, the more attractive. This conclusion also means that an interprovincial tourism attraction is greater than the cross-county tourism, which is consistent with the actual life experience. When the policy is conducive to the flow of population or capital flow, from $\beta \approx \sqrt{T/h^2}$ the T becomes smaller and can lead to the strengthening interaction (Wang 2000).

2.1.4 A Physical Image of Spatial Interaction

Wang (Zheng) gave a statistical physical image of SI in 1986, and later the image was further rigorized, which could be used as a microscopic basis of SI (Wang 2000), which developed the Wolson model to require the system to be closed to protect the system entropy to the maximum assumption.

2.1.4.1 MIF and Population

To recognize the nature of the spatial interaction, we first need to discuss the physical image of spatial interaction. Why is there an interaction between two spatial regions or cities? When populations and funds, etc., move from one region to another region, information exchanges will arise among them: it generates spatial interaction. It shows that the diffusion of a population or fund (etc.) is the microscopic base of the interaction.

In the theory of geographical diffusion, Hagerstrand and his followers have applied the conception of MIF. An MIF is a basic pixel in diffusion; it is not regarded as diffusion, only moves the inner part of an MIF. The conception of MIF is extended. MIF means a least spatial element at which a certain geographical process has a meaning. When a geographical process happened in a certain geographical spatial field, one might divide the field into a great many small fields in a microscope, and show the processes acting in the small area with the same intensity and direction. It is known that the dividing not might carry on to infinity, because of one geographical observer needing a spatial size. This spatial size is just MIF. The MIF is providing with such scale as a small amount h and area size as amount h; in other words, an MIF is a significant minimum pixel in a certain geographical process. MT is often been used in geography. Let us take population diffusion, for example; when the scale that we take into account is only a resident area or a street scale, it is usually meaningless for population diffusion. Another example is that of dividing a region in physical geography; when the width of a "region" attains the size of a plant sample district or a minimum area in which an organism lives, the region is been defined by an observer. A pixel of remote sensing is an FP in a remote sensing image; the characters of pixels condition the image recognization (Wang 1995a). As far as a pixel is concerned, it can only estimate the percent of geographical objects in it rather than demarcate the boundary of those geographical objects again. When it is less than an FP or the length is no less than *h*, it happens unmeasured in the geographical aspect. In a word, FP is the minimum function pixel observed under geography.

With the conception of dimness, we can, even more, introduce the concept of the homogeneous field; in fact, that homogeneous field is naturally extended from the conception of the homogeneous plain of the location theory. In a great many agent studies, geographers have early applied the conclusion of a homogeneous plain to analyze the geographical aspect of non-plain. Therefore, it is clear that geographers have actually applied the concept of a homogeneous field. A homogeneous field is such that a certain kind of geographical process happens in it, such as population diffusion, population succession, location variety, river change, and so on: the effects or disturbance of geographical factors to a geographical process show the same property about a various element of MIF on statistics. In other words, one refers to a field as a homogeneous field in relationship to a certain kind of geographical process. The geographical process in the field either has equal intensity and direction or holds invariable random processes, which have the same probability of distribution.

2.1.4.2 Characteristics of Agent-Based (Population) Diffusion

In a model of geographical diffusion it is generally supposed as physical particle diffusion, but a real geographical particle possesses characteristics different from a physical particle. A monomer of population only has finite longevity, not more than a physical particle. Each people has only finite life; sediment particles in a river are as such yet complete a life course from being started to sinking, and a fund is also as such, which completes its own life in investment. We refer to these geographical particles as an *agent*, for example, a *population*. A population has such features as follows:

1. A monomer of agent has only a limited life in longevity; the longevity is satisfied by various kinds of probability distribution.
2. An agent is a random particle. Although a monomer of agent has a separately migrating object and life meaning, they possess the same nature under a certain geographical process; in other words, they are identical particles about a certain geographical observation. It is significant for us to study a geographical process

only when it regards a single group. For example, if a tourist is a particle, its life starting from becoming a tourist to its end is separate. However, what we study in tourist behaviors is not to study personal behaviors, but to study behaviors of certain group, being the same kinds of person: we do not distinguish one person from the others. This character is a result of that agent as a physical particle.

3. A movement of agent is a spatial process; this action process is called geographical diffusion, because a great amount of agent movement occurs at the same time so that the agents function as a particle with diffusion action on statistics.

2.1.4.3 Model of Agent Diffusion on a Homogeneous Field

With the physical image, one might establish a model agent on a homogeneous field. Supposing these are sufficiently many resident places on a homogeneous field, starting from any resident place a; the agent can reach any of the resident places with finite distance during a finite time, including a return to the resident's own place. Letting the agent set be the whole agent who are born in the resident place a as A, $A_t \subset A$; members of A_t are the agent group who was born in the same time within $[t, t+\Delta]$. Because the field is homogeneous, the probability property of any agent group in migration to any resident place is the same; in other words, the action of agent particle α is two-dimensional Brownian movement. Two-dimensional Brownian movement has been sufficiently studied in statistical physics; as far as a particle that starts Brownian movement from the same time $\tau$is concerned, its probability density function in the spatial sense is

$$f(r,t) = \frac{1}{2\pi D(t-\tau)} \exp\left(-\frac{r^2}{2D(t-\tau)}\right) \qquad (2.1.19)$$

where r is the distance to place a, t is the age of agent groupA_t, and D is the diffusion coefficient, which is defined as

$$D = \frac{h^2}{2T} \qquad (2.1.20)$$

where h is the diameter of FP, h^2 is the area scale of FP, T is the time in which an agent particle from an FP spreads to another next FP, or is the time of agent particle delay in a FP; if we ignore journey time, we can call T the mean migration time. Toward the agent, the diffusion movement begins at life starting at$\tau=0$; we obtain

$$f(r,t) = \frac{1}{2\pi Dt} \exp\left(-\frac{r^2}{2Dt}\right) \qquad (2.1.21)$$

Insofar as Brownian movement of the agent in a homogeneous field is concerned, it is an ergodic process. So, formula (2.1.21) may be regarded as the spatial distribution of probability of members of the agent at time t, appearing in the place where is far from a center place a as distance as r.

Supposing the migration rate of agent group *A* is *g (t)*, the spatial aspect of migratory agent is

$$W(r,t) = Kg(t)f(r,t) \tag{2.1.22}$$

K in the formula is a unitary constant. If *g (t)* is constant, the agent life will be infinite, and will result in $W(r, t) \sim f(r, t)$. However, the agent has life and migration customs, $g\ (t) \neq$constant, and thereby it is necessary for us to estimate *g* (*t*).

We know that the out-migratory rate of an agent has such features: when $t \to 0$, $g(t) \to 0$, that is, it is impossible for a baby to migrate, when $t \to \infty$, $g(t) \to 0$, it is impossible for death to take part in migration. Therefore, let us suppose

$$g(t)\tilde{\ } t^{\frac{1}{2}}\exp(-at) \tag{2.1.23}$$

There *a* is a parameter depending on geographical conditions such as migratory policy, economy, and so forth during the agent process. The relationship (2.1.23) conforms to the fact that the migration is been observed: 0 years old will not migrate, and then the probability of migration increases, and after a certain migration peak, the migration tendency becomes weaker.[1]

Up to now, we can also determine the maximum out-migratory rate age from formula (2.1.23):

$$t_{\max} = \frac{1}{2a} \tag{2.1.24}$$

The existence of the maximum out-migratory rate has been approved by many writers' examples, such as Ding. They found the maximum out-migratory rate equal to about 19–21 years of age in China, which depended on sex, agent policy, and geographical condition, but has nothing to do with the quantity of agent in the residential area directly.

2.1.4.4 The Kernel Spatial Interaction in a Geographical Space

Now let us study spatial interaction. As mentioned earlier, spatial interaction results in agent–particle diffusion, and the agents exchange amounts similarly as photons in electromagnetic interaction. The diffusion of various ages of agent results in a potential *G(r)*, and *r* is the distance far from place *a*. The potential of a field results from interaction, and the interaction force is a negative value of the derivative of potential about the spatial coordinate. We first solve potential function *G*(*r*):

[1]In fact, Poyers et al. (1979) obtained a population migration curve that has this feature. The authors fit the population migration characteristics of locked areas in the 1980s, and similar results were obtained.

$$G(r) = \int_{0}^{\infty} W(r,t)dt \tag{2.1.25}$$

In accordance with the definition of *W(r, t)*, *G(r)* is actually a boundary probability distributed density function of the agent; one can acquire it after *g*(*t*) is to integral. Substituting formulae (2.1.21), (2.1.22), and (2.1.23) into (2.1.25), we can acquire

$$G(r) = \int_{0}^{\infty} \frac{K}{2\pi D\sqrt{t}} \exp\left[-\left(at + \frac{r^2}{2Dt}\right)\right] dt.. \tag{2.1.26}$$

Seeing integral (2.1.26) in the appendix, we have proved:

$$G(r) = \frac{k}{2\sqrt{\pi D^2}} \exp\left(-\sqrt{\frac{2a}{D}} r\right) \tag{2.1.27}$$

This can be abbreviated as

$$G(r) = A\exp(-\beta \bullet r) \tag{2.1.28}$$

Here A is a constant; formula (2.1.28) is the kernel of geographical spatial interaction (SI). We called it the Wilsonian kernel to mark the work of A.E. Wilson who discovered geographical spatial interaction as an exponent function in 1970. Roy and Thill (2003) emphasize that the heart of SI modeling, with some of the deterministic models of micro-economics, has been become probabilistic.

Wang et al. (2004) based a type of spatial interaction on data analysis: knowledge spatial spillover could be expressed as follows:

$$s_{ij} = \alpha_i e^{-\left(\frac{1}{\delta_i^{(k)}} G_{ij}^{(k)} - \mu_i\right)^2 + \sum_{m=0}^{M-1} \frac{1}{\delta_i^{(m)}} G_{ij}^{(m)} - \beta r_{ij}} \tag{2.1.31}$$

where S_{ij} is knowledge spillover strength from region j to region i, δ_i is learning capacity, r_{ij} is spatial distance between regional i and region j, $G_{ij}^{(m)}$ is the mth demand gaps between i and region j region, and M is regional parameters. This model shows that the spatial interaction (SI) is satisfied (2.1.28).

Wang (2000) shows that the exponential form of the spatial interaction is required for a certain condition. In the form of Brownian motion and (2.1.19), whether the agent is population, capital, or knowledge, so long as they have a strong and declining form of life, SI has an exponential form. Further, by the space potential theory (Wang Zikun 1983), we can learn that Newton potential is the result of three-dimensional space Brownian movement; the regional space is also two dimensional; and the Newton potential is not reasonable.

There is an important theoretical achievement, the axiomatic theory of Smith (1987) in theoretical study on the interaction. Set space flow is Poisson flow. Under such condition, each stream can generate a frequency. The mathematical expectation of the frequency of contacting the two places is defined as the interaction. Thus, he extends his theory. In the Smith theory, the spatial generalized distance r_{ij} may be a vector $r_{ij} = \left[r_{ij}^{(1)}, r_{ij}^{(2)}, \ldots, r_{ij}^{(t)}\right]^{\tau}$. $r_{ij}^{(t)}$ is the space geometric distance and communication distance, etc. Correspondingly, β is line array$(\beta_1, \beta_2, \cdots \beta_l)$. $\beta \bullet r$ is scalar quantity. Therefore, the spatial interaction has been generalized.

2.2 Location and Spatial Structure

2.2.1 Lösch Formula

Location, simply speaking, is the space position. The concept of geographical location refers to the location of the enterprise or industry. This position is statistically significant, for example, the agricultural location belt, namely, the profit of agricultural management, in the observation, in which the probability of agriculture is the largest possible to obtain the maximum profit. Observations can been found in this area that there may be other industrial activities, so this area is the recognition of the results of the statistics. On the other hand, location means the most suitable enterprise or facility location. The statistical significance of location here means that this is an appropriate position of economic interests, but not necessarily suitable for engineering, so location analysis is not a project site analysis. The former is the economic significance and the latter is technical. From an economic perspective, the location process is a process of microeconomic activities, whereas from a statistical perspective, it is the macro-representation of the spatial distribution of the enterprises themselves. The particular emphasis is that we usually call the spatial position of geographical things "location," if we consider this in not only the economic sense.

Location theory originated in the Du (Von Thunen 1826) agricultural location theory. With the development of Weber, Hetelling, and Palander, it contributed to Loesch.[2] Lösch (1954) developed the theory of location rent. It defines R as economic profit. x_{ij} is j position from the enterprise. E_i is the production of enterprises in each production cycle. P_i is market price per unit product. a_i is production cost per unit product i. f_i is unit mileage shipping of per unit product. Losch proposed:

[2]See M. Birkin, A.G. Wilson, Industrial location models: 1: a review and an integrating framework, Environment and Planning A 1986; and M. Birkin, A.G. Wilson, Industrial location models 2: Weber, Palander, Hotelling, and extensions within a new framework. Environment and Planning A, 1986 (environment–and -planning.com.).

$$R_{ij} = E_i(P_i - a_i) - E_i f_i x_{ij} \tag{2.2.1}$$

When P, a, f have nothing to do with the production of E, there is

$$r_{ij} = \frac{R_{ij}}{E_i} = P_i - a_i - f_i x_{ij} \tag{2.2.2}$$

$f_i x_{ij}$ also can be expressed as generalized distance C_{ij}. The Losch formula can be written as

$$R_{ij} = \left(P_i - a_i - C_{ij}\right)E_i \tag{2.2.3}$$

This model characterizes the interests of enterprises in the enterprise's interests. The Lösch formula is the basic formula of location analysis.

2.2.2 Thünen Location and Thünen Structure

Thünen imagines that the geographical space is a closed area, which is homogeneous and equivalently fertile, and its center is a central city or place. It can be seen as the market of agricultural products. What kind of regional structure will be formed in space? Where is the location of various industries? It is easy to find by the Lösch formula that space is divided into a circular area around the center. Actually, it is well known by the Lösch center formula that the x-ring from the center has the same interests for a given industry, so the industrial zone must appear in the form of a ring. Assume that the center of the land can be in the form of M. In the center of the $x = D_1$ ring, there is a need to make it more than the other $m-1$. You may wish to note this in the form of l (when there are two types of land use in the form of interest, we are treating it as the form of the same kind). That is to say, within the range of $x < \mathrm{D}$, the probability of using the form l is the largest. According to our understanding of location, this is the location of the industry *1*. With the increase of x, there may be two cases. One is that the land of the product volume has reached the center of the demand. If it continues to produce, it will result in unit profit decline. The second is the increase in freight, which finally makes some of the lower unit prices of product production greater than the form of l products. So, land use form is transformed into that form. By analogy, space is divided into annular space, which is the famous Du regional structure. We can learn from the Liao Shen formula that interest is a linear function of distance, so some forms of use in space can only appear once. In the case of only one center of space, when space is homogeneous, the structure of Fig. 2.1 appears. This appearance is typical of the space structure of the Thünen, and its ring is called the Thünen circle.

It is virtually impossible to only have one space center, and space cannot be homogeneous. Wilson and Brikin (1987) proposed a model to analyze the spatial structure of space. Z_j^k is the land of place j used by industry k. D_j^k is total revenue

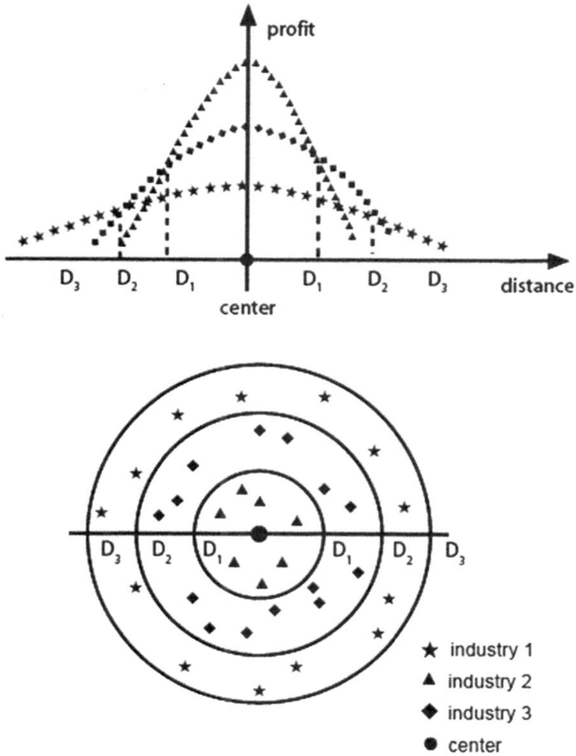

Fig. 2.1 Thünen location schematic form

from the sale of k product. C_j^k is total production cost (including cost, rent, land tax, and freight). Then, the producer's profit E_j^k is

$$E_j^k = p_j^k - C_j^k \tag{2.2.4}$$

Occupancy of the land Z_j^k is set to the monotone increasing function of the profit E_j^k.

$$Z_j^k = f\left(E_j^k\right) \tag{2.2.5}$$

By the Liao Shen formula, the profit of industrial k in the place i (i.e., the center of the i) is

$$r_{ij} = \left(P_j^k - C_{ij} - a_j^k\right)T_{ij}^k \tag{2.2.6}$$

T_{ij}^k is probability of product that appears in i, namely, interaction measure. Total profit of product k of place j is

$$R_j^k = \sum \left(P_j^k - C_{ij} - a_j^k\right) T_{ij}^k \tag{2.2.7}$$

Set the proportion of product k of place j to product k of place $i\, T_{ij}^k$ is proportional to the amount of land occupied by the production of k in place j. There is

$$T_{ij}^k = A_i^k D_i^k Z_j^k \exp\left(-\beta^k C_{ij}^k\right) \tag{2.2.8}$$

Here Z_j is equivalent to O_j of the last section. Superscript k is that the interaction parameter is relevant to the type of products. A_i^k is normalized parameters.

$$A_i^k = \left[\sum\nolimits_j Z_j^k \exp\left(-\beta^k C_{ij}^k\right)\right]^{-1} \tag{2.2.9}$$

Companies follow to maximize profits in the use of land. Therefore, Wilson and Brikin (1987) recommended that the following equation is proposed to describe the process of land conversion:

$$Z_j^k(t+1) = Z_j^k(t) + \sum_{s \neq k} d_j^{sk}(t) \varepsilon_1^s q^s Z_j^s(t) / q^u \tag{2.2.10}$$

ε_k^s is land conversion rate reflecting two forms of use, k and s. d_j^{sk} is the possibility of transformation. When not considering the inertia of the industry, there is

$$d_j^{sk}(t) = \begin{cases} 1 & \textit{If } E_j^k(t) > E_j^u(t), u \neq k \\ -1 & \textit{If } E_j^s(t) > E_j^u(t), u \neq s, s \neq k \\ 0 & \text{Other} \end{cases} \tag{2.2.11}$$

Here q^k is the reciprocal of the marginal effect of land on industry k. By using this model, Wilson and Brikin consider that rectangular space has nine centers, which are evenly distributed along the diagonal, and the land fertile degree ascends to the lower right corner. Calculated by the model of the space industry (in the form of land use), distribution is shown in Fig. 2.2b. The result of this land use can still be called the Thunen type. Its characteristic is that the market (or service) is discrete, and the resource (or service) is a continuous distribution.

Through Thünen location we can see the geographical structure of the boundary. From the microscopic view, the transition is a probability whereas from the macro perspective it is determined and the actual geographical structure of the boundary is not allowed by the probability of the micro perspective.

In tradition, Thünen location is known as "agricultural location," as if Thünen can only fit for an agricultural problem. Actually, otherwise the structure of the Thunen can be generally found to exist in the regional structure of the local area. Currently, it is identified thus. Local area can be from the city center of the CBD out, followed by five nearby rings: central business district (CBD), central

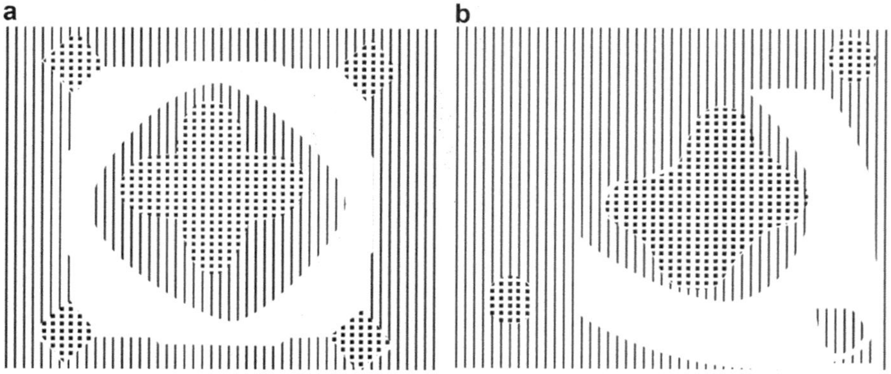

Fig. 2.2 Nine centers of the structure of the Thünen: uniform distribution of resources (**a**) resource linear tilt to one side (**b**) (*Source:* Wilson and Brikin 1987)

fringe area, intermediate residential zone, local fringe zone, and industrial zone. The latter has obvious radial characteristics. At this time, the local area (zone), in addition to the spatial structure, the local area may be a fan (sector). If the country has more than one CBD, it may form a multi-center of the structure of the Thunen. Actually, Wang Zheng (1995a) showed that in a single-center industrial space with the zero-order approximation is cyclic; the first-order approximation is fan shaped.

2.2.3 *Hotelling–Pallant Location and Christaylor Structure*

Hetelling (1929) considered the problem of the different locations. Imagine the market is continuously distributed in the entire space. In space, there are two producers of the same product and the two enterprises should choose their own production location, namely, the location of the enterprise.

Imagine the original location of the two enterprises, A, B. The line segment AB, by the formula we know, is the AB of the vertical split line of the two enterprises (set production costs, unit freight, and purchase price equal) to obtain the same interests. Crossing the line, the B-side, the interests of A are less than B; the class is on the side of the A, and A's interest is less than B's. To compete, let the *a* to *ab* perpendicular bisector direction when a new line is a B′ and a perpendicular bisector, be a new split line in Yuanping junction on the B side. Such *a* captured the market B; to compete, B also moves inward, coincides with the position of the last two enterprises; we commonly split the market space. But this system is not stable, and we may wish to set A to reduce costs, so in the same profit, B product price is higher than A; the result is completely exclusive to B, for B is the only option is to leave A to a new position, the advantage of using distance, to control a market range. However, if the distance is too far again, in the middle of the neutral area a new enterprise appears and competition occurs in a market. So for A, B, the two-enterprise distance that best controls their market junction point is when their profit is zero, and does not allow a third party. In a word, AB is bound to have a certain distance in space, and the system is stable. To compete and prevent the market in the space to maintain a certain distance, enterprises need to determine their own production location, to control and tend to share the market process known as the Hotelling process.

On this basis, Palander (1935) proposed his market location theory. Palander believes that when the position of the enterprise is fixed, it is in the space to produce a price curve and the farther the distance, the higher the price. Until the price curve is reached, the enterprise loses the market advantage.

The intersection of the price curve is the boundary of the market area. The distribution of enterprises (resources) is dispersed, and the location problem of market distribution can be referred to as the Hotelling–Palander problem, or market location problem.

The Hotelling process directly produces the center of the space structure. Actually, as the enterprise (or local) a, B, and another enterprise or local C and its optimal location form an equilateral triangle with ab; halving the space so that they each account for one third. The boundary of the market location is for each side of the intersection. As shown in Fig. 2.3, when each enterprise (center) has the same level of production, the market price is not the result of the origin and change; the space (market) is uniform, the structure of the hexagonal structure, by the Losch formula in the boundary of the interests of neighboring centers.

The center ground structure for Chrisatle (1993) was found during the study of the regional differentiation of the market in southern Germany of so-called enterprise (service, resource) discrete distribution. Regional market structure continuous distribution is the Chrisatle structure. The structure of the center is a special form of the uniform distribution of the enterprise. In the center of the structure, there is a combination of different sub-centers in the center. As shown in Fig. 2.4, $K = 3$ is considered to be suitable for commercial contact because of the role of the center of a higher level of the central role. This finding shows that commodity exchange is

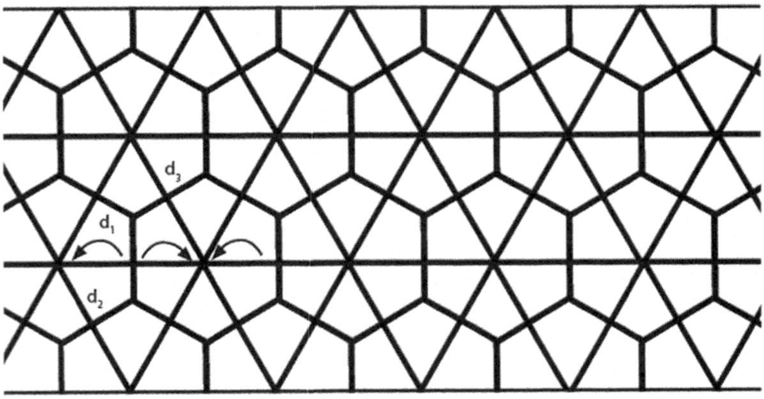

Fig. 2.3 Chrisatler structure formation

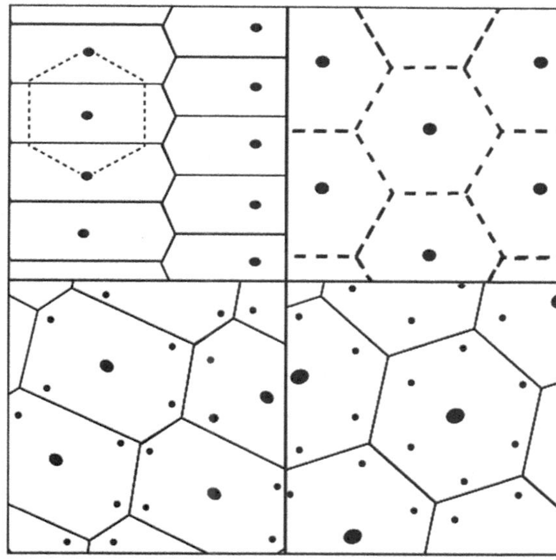

Fig. 2.4 Different forms of symmetrical Chrisatler structures

fully active and free. When $K = 4$, the center is the shortest distance to the center, which is suitable for the transportation of traffic areas or new development areas along the traffic line. When $K = 7$, the central place to the center control is greatest, which is a suitable form of organization of administrative means. When $K = 5$, it can be seen as the transition of $K = 3$ and $K = 7$.

Under normal circumstances, the space has a volatility of economic activity. The new economies of scale x_{k+1} may depend on the previous levels x_k. However, its early development also occupies a part of the market that inhibits the late development. So, in the simplest case

$$x_{k+1} = \mu x_k(1 - x_k) \tag{2.2.12}$$

μ is the parameter. Fluctuations in the level of the economy will also cause the commodity in the market volatility to chase the market. So, the cycle of x_k is the space cycle of economic activity. The latter requires that there be a market or center distribution that is adapted to the period number. As is mathematically known, with the increasing of μ (an increase in the intensity of economic ties) (2.2.12), the state of the type has 2 cycles, 4 cycles, 3 cycles, and 5 periodic solutions. Hence the system of chaos, the corresponding structure number, and the Sarkovskii number of the market cycle. It seems that the order of spatial structure development is 4, 7, 5, 3, 2. If the economic activity continues to strengthen, the spatial structure will appear as chaos. In past years, about 1990, in the southwest of China, economically backward regions often have 7 days in the market habit; more developed areas have 5 days, and the Northeast Yunnan traditional popular market is 3 days. In the developed areas in the South no so-called markets can be seen as examples of space in the process of dynamic change.

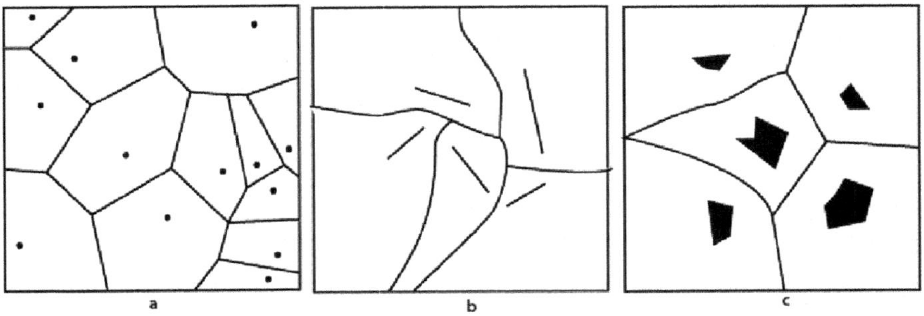

Fig. 2.5 Asymmetrical Chrisatler structure

The center ground structure is an ideal structure. There are two kinds of situations that break the symmetry. One is regional heterogeneity, which causes the development of some local special features, regional and resource heterogeneity, which has been caused by the asymmetrical Taylor Chris structure. Okahe and Suzuki (1987) proposed that the product price is uniform and the freight price is equal in the case of the given set of coordinates. The local structure is a Thlessen polygon. As shown in Fig. 2.5, in the case of a given production of local coordinates$x = (x_1, x_2)$, x belongs to the center of the Tyson polygon conditions. Production costs $P$and freight rates$T\,d_i$ are the lowest.

$$\min J = \sum [P_i + T(d_i)], i \in I \tag{2.2.13}$$

It is the result of space competition to achieve balance. Unit $P_i = P$ (constant), which is in the production cost $C_i = C$ (constant), is to meet the space structure of equilibrium under this condition.

$$\min_{X(I) \in \chi} F(X(I)) = \sum_{i=1}^{n} \int_{V(x_i)} \|x_i - x\| dx, x \in X(I) \tag{2.2.14}$$

$X(I)$ is some kind of space structure. $V(x_i)$ belongs to the control range of the Tyson polygon area or the market range. The meaning of integral $||x_i - x||$is the norm of $x_i - x$. It may be the Euclidean distance or other metrics.

With computer solving, the formula (2.2.14) is still difficult. It requires a structure set within an allowable range for each point associated with the market x to find the possible $x_i - x$definitions of Thiessen polygons. Thus, we can obtain the smallest F(X(I)). A simple method is commonly used in hydrology. Link to the site x_i and draw the vertical line of the connecting line to obtain a division of the space for the centerx_i. The geometric method of this kind of split share market is completely consistent with the (2.2.14) method, but this method still cannot give the structure identification of C and B in Fig. 2.5.

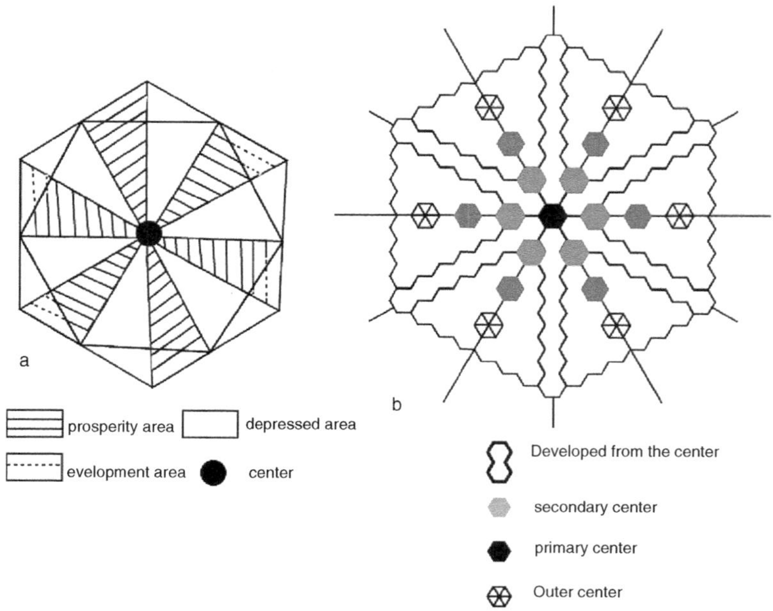

Fig. 2.6 Chrisatler structure diversification. (**a**) Radiating sector. (**b**) Taylor Chris structure

In the foregoing analysis, we always assume that the center is the same level. If the center is of different levels, the spatial structure may have a lower-order structure. In many economic geography textbooks, we can see the diagram of this hierarchy.

Chrisatler structure is also common in urban areas: at this time it may have two kinds of subtypes as shown in Fig. 2.6. In Fig. 2.6a, the local area is divided into a radiating sector, and some branches are developed in particular. In Fig. 2.6b, the local area forms different levels of Taylor Chris structure. In fact, the regional structure is a combination of several ideal spatial patterns. Figure 2.6 shows that the Chrisatler structure is diversification. This diversity or spatial evolution spontaneously occurs because, or because the space is primitive, only the distribution is uneven. In any case, in the road network to bring the convenience of transport, the Chrisatler structure is the cause of diversification.

2.2.4 *Weber Problem and Weber Structure*

1. Weber location

Weber (1909) developed what he called the industrial location theory. The core of Weber theory is the isolated regional system. Weber proposed to find an optimal location in the region to make the enterprise located in the best position of the supply of resources and supply of the market, with the production and transportation costs minimized so as to obtain the maximum profit. Webb believes that the factors that determine the location include general location factors (such as labor costs), special factors (such as water quality), local factors (local conditions), aggregation factors (such as collaborative needs), and natural social factors. In essence, the problem is that the distribution of resources and the market jointly determine the location.

Modern location theory can determine the location of Weber by the Losch principle. Production costs (excluding raw materials, etc.) are fixed. The optimal position of the enterprise is determined by the following formula:

$$\min_{j \in L} \left(\sum_{i,m,n} \left(Y_{ij,m} c_{ij,m} + Z_{ij,n} c_{ij,n} \right) \right) \tag{2.2.15}$$

J is specific business location. L is possible location set. $Y_{ij,m}$ is demand for m products at i points. $Z_{ij,n}$ is supply of resource n at point $i.c_{ij,m}$,$c_{ij,n}$ are tariffs for products and resources, respectively. Generally speaking, there is a conversion relationship between resources and products in the Leontief technology, $Y_{ij,m} = w_{mn} Z_{ij,n}$. So, formula (2.2.15) changes to

$$\min_{j \in L} \left[\sum_{i,m,n} \left(w_{mn} c_{ij,m} + c_{ij,n} \right) Z_{ij,m} \right] \tag{2.2.15a}$$

Or, $x_{ij,k}$ is the (generalized) resource k of i. It includes the market demand and supply of resources in two categories. Then

$$\min_{j \in L} \sum_{i.k} x_{ij,k} c_{ij,k} \tag{2.2.15b}$$

By (2.2.15) we can determine the appropriate location of the enterprise or find the location of the enterprise x_j^*. $x_{i,m}$ is the weight of the resources factor to determine the location of production. $c_{ij,m}$ is transport factor weight. $w_n = \sum_m w_{mn}$ is the weight of raw material resources for product n. There is only one supply point on each theoretical resource. n is the resource point mark. Weber uses the $w(n) = w_n + 1$ to describe the location effect of n, which is called location right. The formula (2.2.15) determines the optimal criterion for the Weber's process. In fact, the criterion determined by the formula (2.2.13) is often insufficient to analyze practical problems. Because the resources are always limited, the resources are bound to be limited by some of the total, and the Webb location process may depend on this restriction. The specific results are shown in Fig. 2.7.

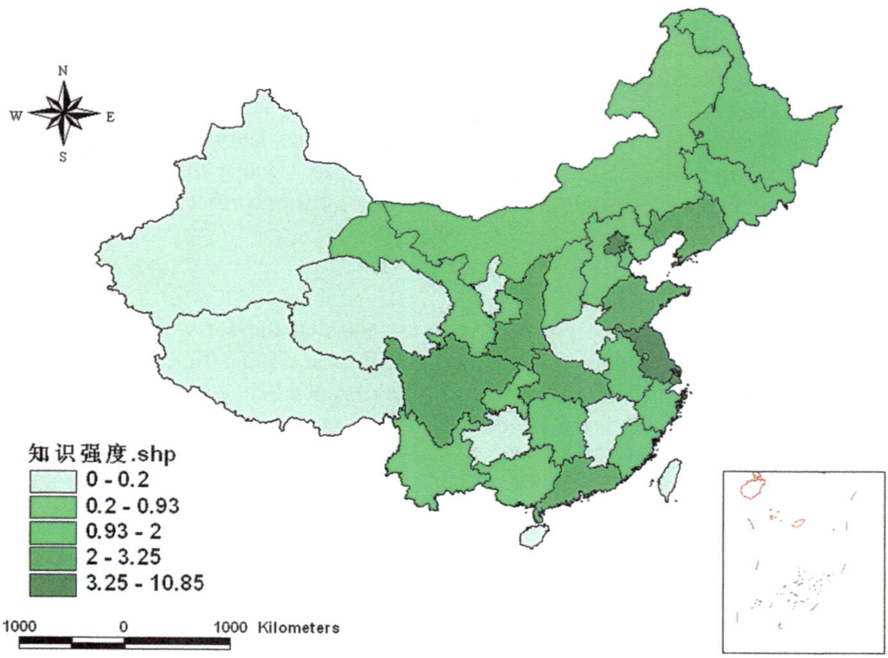

Fig. 2.7 2000 time China knowledge intensity distribution map

2. Weber Structure and Facility location

The Weber location process also produces a spatial structure. A Weber structure is a discrete P enterprise or service to the discrete N market or the allocation of local demand. For example, E_i is the total demand of the market i. (X_i, Y_i) is the market position and is known. (x_j, y_j, w_j) is store location and its business scale. Such a spatial structure is a service or supply or demand problem of the service or demand. The point of this configuration is to promote the facility to be generally, and this creates the facility location problem (Drezner 1994).

The facility location problem is also called the generalized Webb location problem, which emphasizes the relationship between service enterprises, institutions, or facilities and needs. These aspects provide services, collectively known as facilities. Facility location theory is to measure the location of the pros and cons from the point of view of specific facilities. Study of the location of modern facilities had already begun to develop at the end of the eighteenth century and early nineteenth century. In 1990, years later, the facility location problem has aroused the interest of many scholars; all kinds of facilities location problems are put forward to study and discuss. The facilities location problem can be the layout of the facilities into two categories: desired facilities and undesired facilities. The so-called expected facility is to provide facilities for people's production and the people's living environment so that the human environment will not cause harmful effects. This kind of facilities includes hospitals, libraries, schools, markets, fire stations, and other structures. The type of disliked facility is that needed to be used in life but at the same time that which will have an impact on people's living environment, including sewage treatment plants, landfills, and nuclear power plants. In view of these two kinds of facilities, the researchers put forward a variety of criteria to meet the requirements of different facilities layout, and the complete configuration to form a certain geographical structure. Usually this kind of facility needs to configure P; in such theory, the facility location problem can be simply classified into two categories (Drezner 1994).

(a) Median problems

The median point to the problem is to ask the demand for weighted distance services and the minimum, which is equivalent to seeking articles on which we focus. It is required to ensure the needs of the distance between the point and the minimum service facility is the largest global welfare. Its mathematical formulation is as follows:

$$\min C(x) = \sum \{w(x)d(v,x) : v \in V_p\} \tag{2.2.16}$$

$w(x)$ is the heavy weight of each demand point, $d(v,x)$ is the distance from the facility point v to demand point x, V_p is the collection of facility points, and subscript p is facility points and positive integer. The need to solve the problem is to find out the location of the p facility in space and supply chain. Here, order min $(C(x)).w(x)d(v,x)$ may have a more specific mathematical form. This kind of model can be used for layout of public service facilities such as hospitals, markets, or libraries, so here it is also called image forms. Because of the number of facilities, the layout of the facility location is an integer programming problem.

(b) Central problems

The central problem is that we need to point to the demand for the maximum distance or minimum or less than a certain threshold. At the requirement of the maximum distance of the point to the point of facilities, the mathematical expression is as follows:

$$\min g^*(x) = \max\{w(x)d(v,x) : v \in V_p\} \tag{2.2.17}$$

$w(x)$ is weight of demand point x. $d(v,x)$ is the distance from the demand point to facilities. The distance can be rectangular, or the Euclidean distance is the shortest distance on the network. V_p is a collection of facility points. The need to solve the problem is $\min(g^*(x))$. This kind of criterion is usually used for emergency facilities layout, so when an emergency situation occurs the service can be rushed to the scene in the shortest time. Police stations, fire stations, and pipeline repair centers are all this type of facility. So, this kind of problem is called the fire station. In solving the algorithm, the same need applies for an integer programming algorithm.

Often, when a service is not a P, the problem is called the p center of gravity and the p center. The computational complexity of this kind of facility location is related to computational complexity. On this aspect, more can be found in works by Wang Zheng, Wu Jing, and others on "computational geography."

As a specific case, we introduce a specific gravity model. Williams, Kim, and Martin (1990) proposed two criteria to determine the optimal spatial structure of store layout. $Z(N)$ is the objective function. Criterion 1 is the maximum welfare criterion. Even though the buyer to the shop is the smallest, the equivalent form is the combination of the space interaction.

$$\min_j Z(N) = \frac{1}{\beta}\sum_i E_i \ln\left[\sum_j w_j^\alpha \exp(-\beta c_{ij})\right] \tag{2.2.18}$$

The objective function is to be mathematically convenient. (2.2.16) is equal to minimize $\sum E_i\left[w_j^\alpha \exp(-\beta c_{ij})\right]$. w_j is the amount of market resources. The Lagrange function $L = Z(N) + \lambda(W - \sum_j w_j)$. $W - \sum_j w_j$ reflects the resource constraint $W = \sum_j w_j$. It is easy to find the optimal structure.

$$u_j^* = \sum_i \frac{E_i p_{ij} u_i}{c_{ij}} \Big/ \sum_i \frac{E_i p_{ij}}{c_{ij}} \tag{2.2.19}$$

$$w_j^* = W \sum_i E_i p_{ij} / \sum_i E_i \tag{2.2.20}$$

u_i is x_i or y_i.

$$p_{ij} = \frac{w_j^\alpha \exp(-\beta c_{ij})}{\sum_j w_j \exp(-\beta c_{ij})} \tag{2.2.21}$$

Obviously, the city can also be the primary and secondary schools. w is the students' number of x and y is the need for school facilities.

In the foregoing example, we see that the Webb structure does not produce a special spatial pattern, which is because the Webb problem in the market point (point) is discrete, the location is uncertain, and there is no simple space division. Many scholars have been interested in the Webb location problem and its structure is in recent years.

There are some other types of facilities location, one kind being the disliked facilities. For example, a waste disposal site: on one hand, people cannot do without a garbage treatment plant, but on the other hand, they hope to distance themselves from this as much as possible. There is also the definition of facilities in the network space. Requirements and facilities can only be defined on a particular road network. Hierarchy facilities are facilities for different levels of service. Hybrid facilities function not only as a facility but also as a center. Wang Zheng et al. (2010a, b, c, d, e), and Wang Zheng and Peng Youming (2012) discuss these issues.

One of the classical forms of the Weber space structure is Voronoi, which is an outline of space. Its theoretical model is as follows. A set of control points is on the plane, $p = \{p_1, p_2, p_3 \ldots, p_n\}$. Any two points is not a common position, namely, $p_i \neq p_j (i \neq j, 1 \leq i, j \leq n)$ and any four points is not altogether arbitrarily round. The picture shows the Voronoi at any point.

$$T_i = \{x : d(x, p_i) < d(x, p_j) | p_i, p_j \in P, p_i \neq p_j\} \quad (2.2.22)$$

p is Euclidean space in a scattered set of points. d is Euclidean distance. In addition to the outermost points forming an open area, the rest of each point of the Voronoi polygons are formed as convex polygons.

A Voronoi map can actually be seen as a center of space, the heart of Split. It is in accordance with the principle of the nearest distance to each center, the heart of the entire contiguous space split into a number of Voronoi abdominal areas. Each Voronoi region constitutes only one central point. A Voronoi polygon within distance point to point of the Voronoi polygon occurred less than its distance to any other point of the occurrence of the Voronoi polygon. Referred to here is the central point in the Voronoi diagram that defines the set of points, such as the central city.

2.2.5 *Lösch Competition and Lösch Structure*

Space is arranged with a plurality of enterprises, resources, and markets in the spatial continuous distribution. We need to consider resources and market together with enterprises (Center) effect, when discussing the spatial structure. The problem here and with Thunen to different structures is that the latter only considers the interaction of center of resources. This is different from the structure of Chris Taylor. The latter only considers the interaction of market and business center; it is similar to Weberian structure (but the latter resources and market are discrete). So in discussing this, we call the structure of this kind of resource market corresponding to a center for Lorsh structure.

The formation of Lösch structure is the first in the center of local resources. A region has a number of enterprises (center). $p(x, y)$ is density of population. $r(x, y)$is density of population. $q(x - x_0, y - v_0; w)$ is y the residents of the x to the needs of the enterprise products. w is wage level. $l(x - x_0, y - y_0; w)$ is xy (resource) for the supply function of the center (x_0, y_0) under the given wage level. So we can determine the central place to allow the volume of goods and the possible supply of labor (or resources). As in the Eqs. (2.2.23) and (2.2.24) Q_0 is the scope of market attraction; L_s is potential employment.

$$Q_0 = \iint\limits_{\Omega_0} Mq(x - x_0, y - y_0; w)p(r, y)dxdy \tag{2.2.23}$$

$$L_s = \iint\limits_{R_s} SL(x - x_0, y - y_0; w)r(x, y)dxdy \tag{2.2.24}$$

M is total population. S is total resources. Q_0 is the scope of market attraction. R_s is the scope of resource attraction. To labor resources w is wage and can also be understood as a more general resource price. Without loss of generality, the following is a discussion of labor resources. q may be some form of interaction $\exp(-\beta r)$, $r = [(x - x_0)^2 + (y - y_0)^2]^{1/2}$. The Losch form is

$$q = \begin{cases} b - a - cr & \text{while} > a + cr \\ 0 & \text{other} \end{cases} \tag{2.2.25}$$

b is market commodity price. a is production cost, which may be a function of the size and position of the enterprise. c is unit distance commodity unit price. Completely similar, labor supply $l(x, y)$ may only have some form of $\exp(-\beta r)$ or Losch form. It can be taken as a special case:

$$l(x - x_0, y - y_0; w) = \begin{cases} s_0 & if\ w - kr > w_0 \\ 0 & if\ w - kr \le w_0 \end{cases} \tag{2.2.26}$$

s_0 is constant. w_0 is workers' average wage. k is unit single labor commute. Obviously, this is a very simple situation. Ω_0, R_S is $-\infty$, ∞ and we can simplify the integral.

After solving the Q_0, we may, in turn, seek the labor demand in the center (x_0, y_0). Then we reversely introduce the labor demand range R_0, which is a corporate perspective. In turn, we can also introduce the necessary production level Q_s from potential employment L_s and then estimate the occupied market scope Ω_0, which is a government perspective. It is easy to find that when given the $\{\Omega_0(i)\}$, it can produce the $\{R_0(i)\}$. Space is divided by the $\{\Omega_0(i)\}$ of the market and the $\{R_s(i)\}$ of the resource to form the spatial structure $\{\Omega_0, R_0; w\}$. Here, $\Omega_0(i)$ is the market attraction of a central place of i. $R_0(i)$ is the need for the local resources in the center of the i. In turn, when given the resource supply R_s, we can form the specific $\{\Omega_s, R_s; w\}$. Here, the spatial structures $\{\Omega_0, R_0; w\}$ and $\{\Omega_s, R_s; w\}$ are independent and rely on "wages" w_0 – resource price. There is no reason to think that $\Omega_0(i)$ and $R_0(i)$ will coincide the same as $\Omega_s(i)$ and $R_s(i)$.

To determine the R_0 by Ω_0, first we need to determine the relationship between resources or labor and product volume. Common practice is to take the output as a resource of the Douglas function or linear function. That is:

$$Q = \beta K^{\alpha} L^{1-\alpha} \tag{2.2.27}$$

or

$$Q = \gamma L \tag{2.2.28}$$

K is capital stock. When the labor supply is insufficient and the capital is full, K/L can hold a constant so as to make (2.2.27) to linear function form of (2.2.28), which is suitable for determining R_0 by Ω_0.

Usually the space structure $\{\Omega_0, R_0; w\}$ can be divided into three steps: in total space solve Q_0 by the formula (2.2.23); solve L by linear production function (2.2.28) with the Q_D alternative Q.; solve R_0 by L_0. The calculation formula is

$$\iint\limits_{R_0} Sl(x - x_0.y - y_0, w)r(x, y)dxdy = \frac{1}{\beta}Q \tag{2.2.29}$$

This process can only be used to estimate or approximate an analysis by computer.

In an actual situation, business is always trying to make their own interests to achieve the maximum. Therefore, the Q_0 and L_0 of the enterprise are determined by its profit rather than by the maximum allowable range of demand for good. So, we need to propose a profit function $F(Q, L; w)$ to estimate the profit. Order $\frac{\partial F}{\partial L} = 0$ and(or) $\frac{\partial F}{\partial Q} = 0$ and we can obtain the (Q^*, L^*) of the maximum profit. Then use the similar relationship (2.2.24) and (2.2.25) to obtain the optimal spatial structure $\{\Omega_*, R_*; w\}$. A common task is that given the Q or L, such as maximum market Q_0 or maximum absorption of labor L_0, we can find the optimal $L_0^* \; or \; Q_0^*$. Here the subscript "0" is $L^*$and Q^*, which is respectively under the condition of the Q_0 and L_0. In the case of one-dimensional space, Kohlhase and Ohta (1990) gave an example like this. They take the enterprise (center) as located in $x_0 = 0$. Set

$$q(x; w) = \begin{cases} b_0 - b_1(m + cx) & 0 < x < (b_0 - b_1 m)/b_0 c \\ 0 & \text{Other} \end{cases} \tag{2.2.30}$$

$$F(x; w) = mQ - WL - A \tag{2.2.31}$$

m is unit product market price. b_0, b_1 are constant. A is fixed capital required for production. W is nominal wage. w is W/m, which is equivalent to real wage. First set the market radiusr_0. So there is

$$Q_0 = 2\int_0^{r_0} [b_0 - b_1(m + cx)]dx = 2(b_0 - b_1 m)r_0 - b_1 c r_0^2 \text{ when } m < \left(\frac{b_0}{b_1} - c r_0\right) \quad (2.2.32)$$

or

$$m = \frac{b_0}{b_1} - \frac{c r_0}{2} - \frac{Q_0}{2 b r_0} \quad (2.2.33)$$

Take Q and L to meet (2.2.28) and then take (2.2.33) to substitute (2.2.21) and Q to substitute the γL.

$$F = \left(\frac{b_0}{b_1} - \frac{\gamma r_0}{2} - \frac{cL}{2 b r_0}\right)\gamma L - WL - A \quad (2.2.34)$$

Use the partial derivate F on L and make it equal to zero value to obtain the optimal condition.

$$\left(\frac{b_0}{b_1} - \frac{c r_0}{2} - \frac{\gamma L^*}{b_1 r_0}\right)\gamma = W \quad (2.2.35)$$

Or

$$\left(\frac{b_0}{b_1} - \frac{c r_0}{2} - \frac{\gamma}{b_1 r_0} L^*\right)\gamma / \left(\frac{b_0}{b_1} - \frac{c}{2} r_0 - \frac{\gamma}{2 b_1 r_0} L_0^*\right) = w \quad (2.2.36)$$

When given the L_0^*, the $r_0$is the function of w. When given the w, r_0 is an inversely proportional relationship with L^*. In (2.2.35) or (2.2.36), we can see that enterprises can be adjusted to wage W, unit labor wages w, or unit product market price m to achieve changes in the market price of m control range, $r_{0,}$ thus, providing a competitive market, also known as spatial competition. Nakagome (1986) gave a set of simple spatial analysis results of competition. He found that there is an optimal market radius r_0^* in spatial competition, and it may not exist.

In the basic core of the Hotelling spatial competition process, space competition has also led to the location of conjugated phenomenon. That is, a change in position of the enterprise leads to other changes in the corporate market domain; the best regional companies also need to change. This phenomenon is the interaction location conjugate. A discussion about district conjugate can be found in Wang Zheng et al. (2002) in the "theory of economic geography."

Beckmann proves that whether it is a freely accessible market domain, or the case of a welfare-maximizing market domain in the homogeneous plain, space eventually forms a hexagonal structure. Usually, free access will be more than the benefits of the maximum of manufacturers or settlements (Beckman and Puu 1990).

For the development of homogeneous plain space structure, the book of Beckman and Puu (1990) has some fine elaboration.

2.2.6 Space Network Structure

The distribution of the market and the industry is not only the discrete portion but also the dependence of the network. Similar to the Webb structure, consider the market and industry points are in discrete distribution, and these nodes are connected by the unidirectional flow of the arc, and the introduction of the space network structure $\{\mathcal{D}, \mathrm{P}, \mathbf{F}\}$. $\mathcal{D}$ is market point set. P is industrial allocation. F is flow structure of space, namely the involved arc and its flow.

The space network structure is the constraint of the quality of the goods in the balance of demand, supply, and flow:

$$S_k - D_k + \sum_{a \in A(k)} f_a^k = 0 \tag{2.2.37}$$

S_k is production or supply amount of K. $S_k \geq 0$. D_k is the demand for cargo of K. $D_k \geq 0$. f_a^k is liquidity connecting node k arcs on a. Define inflow to be positive and outflow is negative. Obviously, $0 \leq | f_a^k | \leq u_a$. u_a is the maximum allowable flow of arc a. For a road, it is a two-way communication, which is defined as two arcs, and there may be some restrictions on the total flow of two arcs. In common processing, we think that u_a is sufficiently large so that f_a^k is not limited by the maximum allowable flow.

Under the condition of space competition, the spatial structure may reach two states (Friesz et al. 1988): price balance structure and monopoly structure. For the first structure, if the flow of k on arc a is positive, the price p_l of the goods transported from any point L will tend to the local price p_k. Otherwise, the flow is zero. That is:

$$f_a^k > 0, \quad \text{when } p_l + c_a \rightarrow p_k \tag{2.2.38}$$

$$f_a^k + = 0, \quad \text{when } p_l + c_a > p_k \tag{2.2.39}$$

$f_a^k +$ is the flow of k. p_i is the unit price of i. c_a is unit freight rate of arc a. (2.2.38) is a macro situation in that the price is balanced. Actually, when $p_l + c_a \geq p_k$, it can also have goods flow. From a macroeconomic point of view, long-term competition can make $p_l + c_a \rightarrow p_k$. This structure is a large number of enterprises to participate in the competition to obtain the space structure.

When economic activity is monopolized, the Cornel–Nash monopoly equilibrium structure is adapted to the interests of the enterprise. For a monopoly enterprise collection L, its objective function is

$$\max_{f_{a,ij}f_{0,i}} G\left(f_{a,ij}, D_k, S_k\right) = \sum_{k \in K}\left(p_k \sum_{l \in L} D_k^l\right) - \sum_{l \in L} v_l(S_l) - \sum_{a \in A}\left(C_a \sum_{l \in L} |f_{a,l}|\right) \quad (2.2.40)$$

p_k is unit price of k. K is total demand of k. D_k^l is demand of K for L product. L is total supply of l. $v_l(S_l)$ is enterprise cost of l when the production achieves S_l.$f_{a,l}$ is the flow of the good of l on arc a. Then (2.2.39) can convert the quantity

$$S_k^l - D_k^l + \sum f_{a,l} = 0 k = \overline{1,n}, l = \overline{1,m} \quad (2.2.41)$$

S_k^l is the supply amounts of l to k. $\sum_k S_k^l = S_l$. m is supply points. The computational problem of (2.2.38), (2.2.39), (2.2.40) and (2.2.41) is tedious and can be solved by computer.

The most basic form of the space network structure is determined by the type (2.2.37), which is a network flow assignment problem given by the market and the enterprise location. Under the meaning of planning, to people can be in the condition (2.2.37), and the objective function, the formation of other space configuration.

The space configuration structure of the road network, which is based on the road network, is a network problem of microeconomic geography. In macroeconomic geography, not seeking business interests, we may form a hub–network configuration of the structure: the hub has a fundamental significance. This point is not discussed here; see the monograph "Computational geography" by Wang Zheng and Wu Jing et al.

2.3 Other Location Problems

2.3.1 Tourism Location Model

At present, there are three main types of regional development: Thunen's location theory, Weber's location theory, and Parland's location theory.

Thunen's location considers that the resources are continuous; the market is located in the middle of resources (point distribution or discrete distribution); and resources cannot be moved. However, the products can be transported to the central market. Because of the lack of moving, this model cannot be used to study the problem of tourism location.

Weber's location (1909) resources and products can be mobile; the resources can be alternative; resources and markets were tested for punctate distribution. The consideration is how to choose the business location to minimize the shipping freight of resources and products. Tourism resources and products are different in that these cannot be mobile. There is a large difference among tourism resources and there cannot be an alternative. So, the location of tourism cannot be studied by Weber location.

In Parland's location theory, the market is continuous. A number of resources are discrete and can move. The problem is how to divide the market. Because the tourism resources are not available, the location of tourism cannot be studies in the area of Parland. Location is a statistical phenomenon, the formation of the spatial structure of the process with fractal (self-nested) form.[3] Similar to the Chris Taylor (Wang Zheng and Ding Jinho 1994) structure, the high grade of the structure contains low levels of structure. Two kinds of structure have great similarity. Tourism location should also have such characteristics.

However, because of the unique tourism location, the typical location theory cannot be fully applied. The unique features of tourism location (Wang Zheng and Wang Ying 2000) are, first of all, that tourism resources and tourism products do not have mobility; consumers who want to consume the tourism products must start from the source, through the means of transport to the tourist resources. Second, past economic theory regards products as homogeneous, whereas tourism products have differences; each scenic spot is different, there is a kind of individual difference, even though the two scenic spots belong to one type, there are individual differences, so there is competition and spillover between tourism products. Because economics has just begun to pay attention to this problem, as a branch of economic geography, tourism geography in the past 20 years, in fact, did not receive serious study of location.

1. Regional model of tourism industry

Tourism is a kind of consumption of tourists; tourist attractions are the product of the tourism industry. Tourists choose a point to visit or not to decide on the effectiveness of its feelings. For a tourist spot, the time and the time required to travel to the place where the tourists are choosing whether or not to travel are the biggest determinant. Therefore, Wang Ying and Wang Zheng (2000) defined the tourism utility function as

$$f = T_y/T_x \tag{2.3.1}$$

T_x is the time needed to return to a place from a temporary residence to a tourist destination. T_y is in the time of the tour. Therefore, it is mainly to determine whether a tourist spot has a tourism value, that is, whether the marginal utility of the tourist spots is satisfied $f \geq 1$, namely $T_y \geq T_x$. If in a tourist visit time is greater than to the increased diners to travel time, travel psychology will meet, will go to the tourism, the tourism value.

[3]On the nature of fractal images, refer to Chen Shouji and Zhang Liming (1998).

In the Wang Ying and Wang Zheng model, although simple, the economic structure is not clear. To this end, Wang Ying and Wang Zheng (2000) improved it as follows. Similar to the general economic utility, the utility is Cobb Douglas type. There are two factors u, v.

$$U = u^{\alpha} v^{1-\alpha} \tag{2.3.2}$$

Order $u = T_y, v = \frac{1}{T_x}$, then easily obtain

$$U = T_y^{\alpha} (1/T_x)^{1-\alpha} \tag{2.3.3}$$

α is preference. This is a more general model than the original model. Different people have different preferences, and the quality of tourist destinations differs. The results αmay be different. In general, $\alpha = 1/2$.

$$U = \left(\frac{T_y}{T_x} \right)^{1/2} \tag{2.3.4}$$

It is equivalent to (2.3.1) in the identification of whether the tourism area has industrial prospects. According to this mechanism, Wang Ying and Wang Zheng (2000) and Wang Zheng and Wang Ying (2000) according to their observation of tourist attractions divided these into four bands: the scenic spots, singular scenery, stagnation, and ethnic customs. Where there were historic sites and scenic spots in local and suburban areas, because of people's cultural habits, these zones formed a gathering of places of interest. Farther from the center of the place where it can be set out as about a day trip to the region, it has become a part of the singular landscape found to be used. As a result if the trip cannot be completed within a day of tourism, tourists will choose only those spots famous in the region that have a special landscape status that can support more than a day of time travel or has special educational significance of the special scientific and educational culture; other scenic areas are not visited. Further are the areas where there is a difference in times of natural landscape, local customs and practices, and the central region geography. Here the travel time is generally 2 days or more.

2. Tourism market location model

In the traditional location theory, Palander location (1935) is based on the market. Parland proposed space has an enterprise; its market is much, then consumers can accept the price. Because consumer location in tourist attractions and tourist places in space can be regarded as a discrete (point) in the distribution of resources and market, although the tourism resources do not have mobility, but visitors can through transport reach the area for consumption. In this process we need shipping costs, so the tourism consumption location problem can be as classed as a Paronnaud location problem.

In the market analysis, there are two problems: the space market and the consumer circle. Space market partitioning and market share issues. The consumer circle can measure the main center of the area in the area of the impact of the area, and according to the area of the consumer can determine the area of the main marketing object.

We can calculate the travel consumer circle from two perspectives: these two algorithms are given in accordance with the budget constraints.

A. Angle of view: to analyze the radius of the source of the source area of the customer.

Usually tourist travel expenses include long-distance transportation, accommodation, meals, sightseeing, shopping, and entertainment, but not all visitors will carry out these consumptions. Visitors, on the basis of their own economic situation, will make choices, not for shopping or entertainment, that money is spent on travel-related activities. According to the tourist consumption structure, we can get the corresponding three kinds of consumer circle: entertainment circle, normal circle, and economic circle.

Entertainment: tourists in the circle to a scenic spot in addition to travel, accommodation, shopping, also entertainment; the entertainment circle radius calculation formula is as follows.

$$R_a = \frac{E - L - F - T - S - A - O}{2t} \tag{2.3.5}$$

R_a is travel distance, E is per capita tourism total cost, L is per capita tourist accommodation, F is per capita tourism and catering costs, T is per capita travel expenses, S is per capita travel shopping cost, A is per capita tourism entertainment expenses, O is other expenses (including city traffic, post and telecommunications, other expenses), and t is per capita traffic mileage.

The normal circle: the circle of tourists to the scenic tour, accommodation, shopping, but not for entertainment. So the radius of the normal circle is calculated as (2.3.6). That is, the molecular ratio (2.3.5) is one of the few (per capita tourism entertainment expenses).

$$R_n = \frac{E - L - F - T - S - O}{2t} \tag{2.3.6}$$

Economic circle: the circle of tourists to the scenic tour and only spends on food and rent, not for shopping and entertainment activities. Therefore, the formula for calculating the radius of the economic circle is:

$$R_e = \frac{E - L - F - T - O}{2t} \tag{2.3.7}$$

We have 1999 domestic tourism data: the cost of accommodation accounted for 24.06%, food and beverage costs accounted for 18.65%, travel expenses accounted for 15.28%, shopping expenses accounted for 8.17%, and other expenses accounted for 8.67%, the per capita consumption level of 614.8 yuan (China Statistical Yearbook 2000). By querying the railway time table and odometer, in Shanghai to the average length of Shangrao, the railway traffic fee is $0.2296/miles per capita. When these data are put into the formula of calculation, three tourism circles of radius can be obtained: for entertainment R_a is 195.07 kilometers, for normal ring R_n is 304.45 kilometers, and the economic circle R_e is 509.03 km.

B. Perspective two: to the destination as the center, analysis of the budget constraints of the market area.

In a given destination planning, a common problem is in the range of the scenic spots should be much larger than the consumption structure of visitors, which puts forward the destination for the center of the consumer location problem. According to the number of days of travel and the daily cost of calculation, it is been assumed that the space of tourism consumption is the same. The radius of the zero consumption circle is the total cost of tourism is equal to the total travel distance of the national per capita total travel expenses. Taking into account the stronger consumer spending more money to travel, so put forward the times of the consumer, that is, visitors spend two times the total travel distance. The formula is as follows: (2 is divided by the total distance divided by 2 to get the travel distance).

Zero consumption circle:

Travel distance = ((Per capita tourism total cost − every day to eat, to play, and spend × travel days)/traffic mileage)/2

Times consumption circle:

Travel distance = ((per capita tourism total cost × 2: every day to eat, to play, and spend × travel days)/traffic mileage)/2

In 2000, in China's per capita tourism total cost of 614.8 yuan, traffic mileage fee is $0.2296/km; in the calculation of the lodging, a day to play, or spend according to the actual situation of the fill in, according to the number of days of the tourism can simulate travel radius.

From the concept of the consumer circle, Li and Wang (2009) developed the concept of tourism circle to further improve the tourism geography for many years. They think that the tourism circle is a kind of spatial structure of destination, and the space unit of the space unit in the market needs to form a "center hinterland" space combination. As the tourists may pay the travel time, the radius of the tourism circle is been enlarged, and the tourism circle is only possible to reach a certain degree. (See Li and Wang 2009.)

(c) Tourist traffic location model

Traffic condition is the basic condition of the development of the tourist area, so the traffic location analysis is an important problem in the analysis of tourism planning. Traffic location is a point in the traffic network location; we use the network index to evaluate its convenience. From the abstract concept, the traffic location of tourism is the topology graph, which is composed of the network and the nodes. Nodes are composed of scenic spots and distribution centers, and the network is the traffic line. So it could been used to measure the traffic location in the network.

The concept of network index is as follows (Yang 1987):

Order p = The network number (the number of subgraphs, segmentation of sub-networks in the network is not the number of connections); m = connection number; n = node number. Have the following metrics of network connectivity:

1. α: The ratio of the actual number of rings to the maximum possible number of rings in the network. $\alpha \in [0, 1]$.

   Actual ring number in network $= m - n + p$
   Maximum possible number of rings $= 3(n - 2p) - (n - p)$
   α = the number of actual rings/networks in the network.

For the tourism transportation network, α is the better, the link between the nodes more, and the traffic is more convenient. Generally, $\alpha > 0.5$. The traffic network can be regarded as convenient.

2. β The ratio of the number of connections and the number of nodes in the network. $\beta = \frac{m}{n}$ In graphing the theory has been demonstrated $\beta \in [0, 3]$. When using the index β to measure the tourism transportation network, we think that the greater the β, the traffic routes, the better. Generally, $\beta > 2$. The reachability of network nodes is good.
3. γ. It is the ratio of the number of edges and the maximum possible edges in the network. $\gamma \in [0, 1]$.

   Maximum possible number of edges of the network$= 3(n - 2p)$.

γ is the actual number of edges in the network/maximum possible number of edges of the network. γ is the index of the degree of traffic line density. The bigger the γ, the greater the traffic line is. $\gamma > 0.7$ shows that the tourism transportation line.

4. η. The length ratio of all edges and the number of edges in the network, namely the average length of wire. The longer η shows that the length of the tourism network traffic line, shows that the degree of dispersion between the larger attractions, or the distance between the scenic spots.

2.3.2 Location Factor Model of High-Tech Industrial Agglomeration

The industrial gathering is not just anywhere; it has a certain location. So what is the choice of enterprises in the region to gather it? On the other hand, on the macro scale, in which areas can the industrial gathering form? Under the tide of globalization and technology, regional development is not balanced, some areas seem to have a special "sticky" to attract capital, technology, and foreign enterprises at the local root, to become the center of industrial agglomeration, and thus to experience a boom of development. What is the impact of the location of the enterprise aggregation? From a macro point of view, the industrial clustering location factor model is used to analyze the factors affecting industrial agglomeration.

1. Factors analysis of industrial agglomeration

Knowledge spillover environment: the modern industry is a kind of knowledge-intensive industry; the main reason is that it relies on knowledge spillovers. Knowledge spillover is an economic phenomenon that the new economic growth theory emphasizes. Every step of the enterprise requires a certain knowledge base, including the knowledge of the enterprise production and the "know-how." In addition to the continuous increase in the experience of enterprise employees, the basic knowledge can also be expanded through other external resources (such as the industry's journal, market research, technical exchange meeting, etc.); this behavior will make the corresponding behavior of enterprises in the production value new information to obtain considerable profit, which is knowledge spillovers. Knowledge spillovers have a characteristic, deviate from the source of knowledge; knowledge spillover intensity decreased rapidly (can product Is, Caniëls and Verspagen 2001; Wang et al. 2003a). In this case, to benefit from the knowledge spillover, enterprises will actively be near a source of knowledge or to provide knowledge about the environment.

Human capital intensive: Knowledge spillover is a kind of external factor. It is a kind of human capital to use knowledge and enjoy the benefits of knowledge. According to the classical location theory (Lu 1989; Wang Zheng et al. 2002), the resource is the basic factor in determining the industrial location. For many modern industries, the use of resources is human resources; this kind of resource demand led directly to industry gathering in the region of human resources.

Climate environment: In industrial concentration, only knowledge of the environment is not enough. The Wang Zheng et al. (1998) calculations show that the spatial correlation coefficient of the two is only 0.284 when considering the distribution of universities and high-tech enterprises, and the coefficient is 0.587. For this phenomenon, Wang et al. (1999) proposed that the distribution of high-tech industrial belt has a significant relationship with the regional climate of life, unfortunately, with no further explanation. Here we explain this kind of phenomenon in modern industrial activities depends on R & D activities and on the human comfort climate environment, whether the weather too hot or too cold, or whether high-altitude atmospheric oxygen deficiency will reduce the efficiency of R & D activities, so high-tech enterprises that lack recreational options are in competitive disadvantage that can be attributed to failure. This solution can explain the decline of the United States 128 highways because of relatively poor and regional climate conditions.

Supply chain environment: In general, the final product of the modern industry is complex, requiring many intermediate products, in accordance with the modern industry logistics management theory, which leads to high-tech enterprises needing to survive in a good environment for the supply of the industry. In fact, any one of the production enterprises is in a dual status in the market, relatively speaking; and the enterprise is both producer and consumer. It needs to buy other products as the production input, but also for the production of their own products sold to other enterprises. It is this input–output relationship to the production of multiple enterprises, which is linked to one industrial chain, to achieve the goal of saving the cost of enterprises in the chain, that the choice is to be close to each other.

Traffic environment: From a larger scale, high-tech enterprises need to appear in the area of good traffic conditions, the kind of traffic conditions that protect the supply of smooth information exchange and commodity circulation. Traffic conditions in the Du can understand the classical location of the attention; in the location of the Webb model, traffic conditions are the most basic conditions to consider. In recent years, the new economic geography has raised the cost of transportation, so that traffic conditions must be considered (Fujita et al. 1999). In the high-tech industry location selection, the traffic condition is more prominent than the transportation problem, because the high technology product transportation quantity is reduced, but need is to have fast information exchange and the advantageous knowledge spillover. In fact, knowledge spillover is been accomplished by the diffusion of knowledge, and the advantage of knowledge diffusion can be constructed.

Trade terms: With the development of modern communication technology, the proportion of freight in the cost of goods is getting smaller and smaller, and the distance between the space and the freight is also greatly reduced. Transportation cost is only one form of transaction cost, which includes all aspects of the impact of product circulation, such as taxation, politics, institution, language, culture, and so on. The importance of terms of trade had an important role in the development of Cambridge Industrial Park, and it had a part in the development of the Guangdong high-tech industry. At a deep level, the trade promotes the knowledge spillover of the basic elements of the aggregation.

2. Location factor

First, consider the measurement of knowledge spillovers. The aggregation level of the university can be used as a measure of knowledge spillovers because knowledge spillovers are the first to focus on the creation of knowledge. This intensity reflects the innovation ability of knowledge accumulation, and the degree of the university's aggregation reflects the cultural accumulation and innovation ability of a region. We use the "large" 5-year University Chinese evaluation score estimation of the regional knowledge level of a place. To this end, the top 100 schools, to its score based on the 5 years of the rankings to do an average, in each region of the university to do a cumulative, to get the local knowledge of the environment.

Local human capital is another factor that is highly praised. To calculate the human resources, we applied the model of Jeong (2002) development:

$$\frac{H_s}{H_u} = \frac{Y_s}{Y_u} \frac{w(u, \tilde{u})}{w(s, \tilde{s})} \tag{2.3.8}$$

Consider two areas s and u. Here Y_s and Y_u are respectively the total national output of s and u. H_s and H_u are, respectively, total human capital investment of this two districts. Order $(s, \tilde{s})$ and $(u, \tilde{u})$ are two different people. $w(u, \tilde{u})$ and $w(s, \tilde{s})$ are their wages. They offer the same human capital investment in different regions, the model that a country's total human capital investment is equal to the total labor income divided by the average annual income of the industrial workers. This calculation can be used to compare the status of human capital investment.

In addition, we give the human capital investment according to the years of education. It is said to be human resources, or human capital stock. The workers' per capita education model is as follows:

$$E_t = \frac{\sum_k v_k L_k(t)}{L(t)} k = u, s, j, p, i \tag{2.3.9}$$

Here the subscripts u, s, j, p, i are, respectively, the education level of junior college, high school, junior high school, primary school, and illiteracy. Formula (1) gives the state of the human capital investment, and formula (2) is reflected in the relative quantification of human capital. The ratio of the two is defined as the effective use of human capital. It is been found that the coefficient of the eastern provinces is higher, whereas the middle is in a low efficiency. The specific results are shown in Fig. 2.8.

Another important factor affecting the location of high-tech industries is the climate to the people. We use ICL (De Freitas 1979):

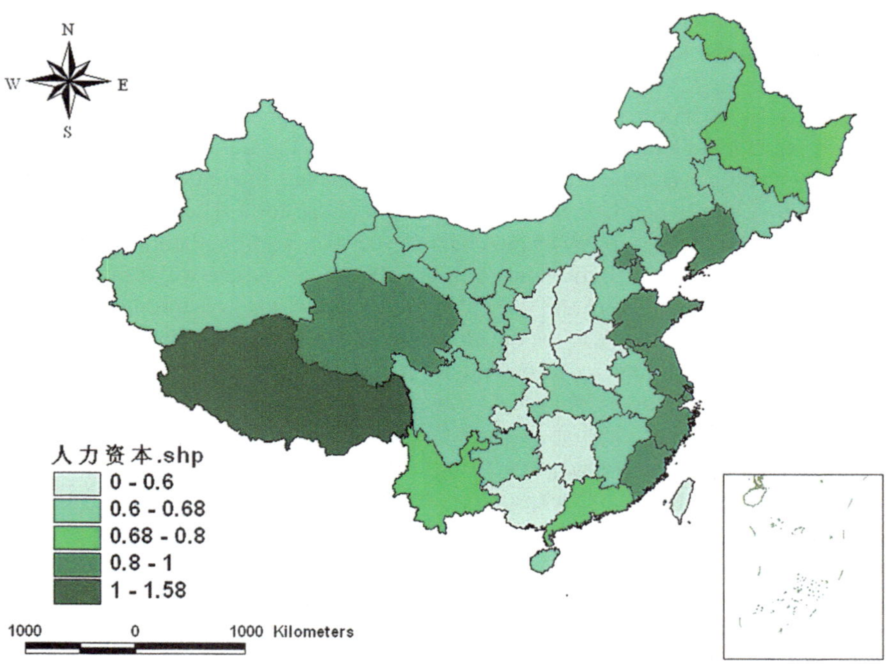

Fig. 2.8 2000 time of the effective distribution of human capital in China

$$\text{ICL} = (33 - T_0)/0.115H - aR\cos\alpha/[(0.62 + 1.9v)H\} \tag{2.3.10}$$

The physical meaning of ICL is the value of thermal damping material, whose unit is OCL. T is temperature, whose unit is centigrade. H is 75% of human metabolic rate. The absorption of sun to the human body is related to body posture, clothing albedo, and the sun's direct radiation, scattering, and reflection components. In most cases, the maximum value is 0.06. R is solar radiation that is received by a unit of land that is perpendicular to the sun: 1367 $\text{wm}^{-2}\,\text{s}^{-1}$. α is solar zenith angle, V is the wind speed. This chapter takes an average of all parts of the country. We define the climate adaptation level of life as a place for the climate conditions most adapted to human life of the deviation DICL, which is the hottest and coldest month from absolute deviation and absolute.

$$\text{DICL} = |\,\text{ICL}_h - 1.0\,| + |\,\text{ICL}_c - 2.0\,| \tag{2.3.11}$$

Here the agreement is between 1 and 2 for the index of the most suitable climate for life.

Through fitting analysis, we found that the knowledge environment, human capital, and climate conditions have basically explained the regional effect of China's high-tech industry, with significant statistical results. For the development of high-tech industry, the emergence of supply chain environment, business environment, and traffic conditions seem to be ignored. We believe that the reasons have two aspects: first, such as our analysis of the supply chain environment, the impact of the business environment in the high-tech industry has a certain size of development; the level of China's high-tech industry development of these factors is not strong.

Finally, it is worth emphasizing specially that important features of spatial process and location theory are a kind of macro theory, the statistical characteristics of its focus on geographical things in the earth's surface movement. For example, for a factory site, when it is located at the river shore for a determined period of time, we consider the stability of the channel, the river section, then in general it is necessary to study the individual "micro". However, when we turn from the industrial location to determine the location of the factory, we introduce a macro view: a region is conducive to the development of the industry, spatial analysis only pointed out that at large range, a specific location needs space, outside the academic analysis. The spatial theory of space is a two-dimensional abstraction of the surface of the earth, so it is necessary to ignore some of the "surface" details, which must be based on statistical characteristics to define the nature of the space. Only the statistical stability of the background to the analysis of the regional process, population process of the "reference system," is important. In short, in geography, spatial analysis is a macro theory, although some space theory is been considered as a "micro-economics" from the perspective of economists, and the geographical concept must be acknowledged.

Appendix

To integration

$$G(r) = \int_0^\infty \frac{k}{2\pi D\sqrt{t}} \exp\left[-\left(at + \frac{x^2}{2Dt}\right)\right] dt \tag{A.1}$$

Let

$$x = \sqrt{at} \tag{A.2}$$

We have
The second integration is a general integration with a known result:

$$dx = \frac{\sqrt{a}}{2\sqrt{t}} dt \tag{A.3}$$

$$G(r) = \int_0^\infty \frac{k}{\pi D\sqrt{a}} \exp\left[-\left(x^2 + \frac{r^2 a}{2Dx^2}\right)\right] dx \tag{A.4}$$

$$\int_0^\infty \exp\left[-\left(x^2 + \frac{b^2}{x^2}\right) dx\right] = \frac{\exp(-2b)}{2} \sqrt{\pi} \tag{A.5}$$

Comparing (A.4) with (A.5), it is easy to obtain

$$b = r\sqrt{\frac{a}{2D}} \tag{A.6}$$

$$G(r) = \frac{k}{2\sqrt{\pi D^2}} \exp\left(-\sqrt{\frac{2a}{D}} r\right) \tag{A.7}$$

or

$$G(r) = \frac{k\beta^2}{8\sqrt{\pi}a^2} \exp(-\beta \cdot r) \tag{A.8}$$

So has
Let $K = \frac{k}{8\sqrt{\pi D^2}}$ We obtain formulas (2.1.27) and (2.1.28).

Chapter 3
Spatial Dynamics

Abstract In this chapter, the author implements the line of understanding the phenomenon of regional science from the point of view of statistical physics, and expounds the spatial dynamics supported by the basic unit of population and capital from the aspect of statistical physics, trying to explain regional science on a dynamical basis. As a development of spatial dynamics analysis method, in this chapter, the cellular automaton and agent-based simulation (ABS) methods are introduced. In particular, the author gives examples of agent-based modeling in regional science in combination with China's regional development.

3.1 Statistical Mechanics

3.1.1 Basic Theory

The basic theory of classical spatial dynamics is the movement of life, which is the basic image of the human body (agent) or the "population", resulting in spatial aggregation and anti-aggregation (disaggregation).

The classical theory of stochastic diffusion theory, by Kanaroglou et al. (1986), is a classical description of spatial dynamics.

L regions exist in the design space, and N is the main body (population, capital, etc.); n is a vector of the distribution of the population.

$$\boldsymbol{n} \equiv (n_1, n_2, \ldots, n_N)^{\tau}, 0 \leq n_i \leq N, \sum n_i = N \tag{3.1.1}$$

Given a determination of the ground state n, the state m, which is defined by the space system, is an offset of the initial state n.

$$\begin{aligned} \boldsymbol{m} &\equiv (m_1, m_2, \ldots, m_N), 0 \leq n_i + m_i \leq N \\ |m_i| &\leq N, \sum m_i = 0, \sum (n_i + m_i) = N \end{aligned} \tag{3.1.2}$$

Set

$\Omega = \{m + n\}$ and $\Lambda = \{m\}$.

Further, some assumptions are needed to improve the theory.

Z. Wang, *Principles of Regional Science*, New Frontiers in Regional Science: Asian Perspectives 15, DOI 10.1007/978-981-10-5367-2_3

Hypothesis 1: the system is immediately dependent on the state of the last observation

$p(n+m; t+\Delta t| n; t)$ is the state of the T system to the state of the N, the probability of the transformation to the states $n + m$. At each moment of the system, only one state exists. So we have

$$p(n+m;t|n;t) = \begin{cases} 1 & \text{when} \, m = 0 \\ 0 & \text{Other} \end{cases} \tag{3.1.3}$$

$$\sum_{m \in \Lambda} P(n+m; t+\Delta t|n; t) = 1 \tag{3.1.4}$$

With the Karl Chapman–Kolmogoroff formula, and

$$P(n; t+\Delta t) = \sum_{m \in \Lambda} P(n; t+\Delta t|n+m; t) P(n+m; t) \tag{3.1.5}$$

$P(n, t+\Delta t)$ is the probability of a system in a state n at $t+\Delta t$. By the assumption one can prove that the space system satisfies the master equation:

$$\frac{d}{dt} P(n,t) = \sum_{\substack{\xi \in \Lambda \\ m \neq 0}} [W(n; n+m) P(n+m, t) - W(n+m; n) P(n; t)] \tag{3.1.6}$$

Here

$$W(n; n+m) = \frac{\partial}{\partial t} p(n; t+\Delta t|n+m; t) \tag{3.1.7}$$

or

$$W(n; n+m) = \lim_{\Delta t \to 0} \frac{p(n, t+\Delta t|n+m, t)}{\Delta t} \tag{3.1.8}$$

Call the probability of the system state from the $n + m$ to the N in the unit time. For the significance of the master equation, we can further refer to statistical mechanics texts.

$p_{ij}(t, \Delta t)$ is [t,t+Δt], the probability of individual particles migrating from area i to area j. The density of the spontaneous transition probability is

$$q_{ij} = \lim_{\Delta t \to 0} \frac{p_{ij}(t, \Delta t)}{\Delta t} \tag{3.1.9}$$

q_{ij} and q_{ij} Δt is in the internal particle system and the probability of transferring from state j to state i, respectively. It describes the transfer of the particle system state rather than the state of the individual particles.

Hypothesis 2: the particle is independent of the movement

With the help of hypothesis 2, statistical physics has been proven to be the main (3.1.6).

$$\frac{d}{dt}\langle n_j(t)\rangle = \sum_i \langle q_{ji}(t)n_i(t)\rangle - \langle q_{ij}(t)n_j(t)\rangle, j = \overline{1,L} \quad (3.1.10)$$

<x> is the statistical equilibrium of random variable x, which is macroscopically measurable. Statistical thermodynamics has demonstrated that when the number of particles is sufficiently large, it can only be detected as the <x> in the macro.

$P_i = \langle n_i\rangle/N$ (3.1.10) is the generation of the order; then, set the time. On behalf of the constant, we can turn (3.1.10) into

$$\frac{dp_j}{dt} = \sum_i \left(f_{ji}P_i(t) - f_{ij}P_j(t)\right), j = \overline{1,L} \quad (3.1.11)$$

Clearly solving the (3.1.11) requires the specific form. From Sect. 2.2 of the previous chapter, we know the spatial interaction with probability. Physical changes in (3.1.11) $P_i(t)$ can be understood as a change in the relative population of I, the cause of the change of population may because of the mechanical process or the migration process. In the case of N, the change P_i is dependent on the migration, so in (3.1.11) the right-hand side indicates the migration process. Naturally, the relative proportion of j to i is explained. $f_{ji}P_i(t)$ is a population of j. $f_{ij}P_j(t)$ is a population outside j. f_{ij} reflects the measure of the migration between i and j. As we discussed in Sect. 2.2, f_{ij} should have the meaning of space interaction. Or, we can agree that it is the intensity of the space interaction. So at least it should be similar to the general spatial interaction $\exp(-\beta C_{ij})$.

So long as the birth rate remains unchanged throughout, the relative number of P_i independent bodies is also maintained, so that to adapt to the analysis of the birth rate is not zero. For (3.1.10), when $q_{ij}=f_{ij}$ (time independent), to analyze population dynamics, it is also on the right side of the birth rate γ. Zhang Weibin proved that in some special circumstances the solution of (3.1.10) and (3.1.11) multiplies by $\exp(\gamma t)$ and is equal to add γ on the right side.

The foregoing discussion is restricted to the case of space as discrete; it cannot be discussed in the regional or local development of the regional expansion, the phenomenon of central migration; for this consideration we need to consider space as continuous.

In the continuous space, the master equation has the following form, called the Fokker–Planck equations or the Kolmogorov equation.

$$\frac{\partial}{\partial t}p(t,x) + \frac{\partial}{\partial x}(a(x,t)p(t,x)) - \frac{1}{2}\frac{\partial^2}{\partial x^2}[b(x,t)p(x,t)] = 0 \quad (3.1.12)$$

$P(t,x)$ is the probability density of x at t, $a(x,t)$ is offset coefficient of random process, and $b(x,t)$ is diffusion coefficient. Generally, (3.1.12) is used to consider the propagation of spatial diffusion.

3.1.2 Spatial Aggregation and Differentiation

1. **Aggregation**

Weidlich and Haag (1987) studied spatial aggregation by using the principle of spatial dynamics: There are L places. n_i is the state, N is the total population of L places, and $W_{ji}(t)$ is migration from local i to j in the unit time $W_{ji}=p_{ji}(t)n_i(t)$. The interaction $p_{ji}(t)$is

$$p_{ji}(t) = W\exp\left[-\kappa C_{ji}(t)\right] \quad (3.1.13)$$

The generalized distance is defined as the difference of the population.

$$C_{ji}(t) = n_i(t) - n_j(t) \quad (3.1.14)$$

The population change rate of the local j is from the main equation:

$$n_j = \sum_{i=1}^{L} Wn_i(t)\exp\left[\kappa\left(n_j(t) - n_i(t)\right)\right] - \sum_{i=1}^{L} Wn_j(t)\exp\left[\kappa\left(n_i(t) - n_j(t)\right)\right] \quad (3.1.15)$$

Solve (3.1.15) where there is a critical value found: $\kappa_c = L/2N$.

1. When $\kappa < \kappa_c$, the homogeneous state $(n_1 = \cdots = n_L)$ is a stable equilibrium state. In the region, there is a tendency to be homogeneous, and there will be no space to gather. When $\kappa > \kappa_c$, even though the uniform state is the equilibrium point, it is unstable and will appear in the space to gather.
2. When $\kappa_d < \kappa < \infty$, a place to grow, L^{-1} is a place where the recession of autonomy to the central place continues to focus on the emergence of the center of the "occasional" spatial fluctuations; the uniform state is not a stable equilibrium point. $\kappa_d < \kappa_c$ is a new threshold.
3. When $\kappa_c < \kappa$, there is a new equilibrium point. In this process there are P = L/2 places where there is a dense P where there is a dilute self-subject. The density difference of the main body is determined by the aggregation parameters; when$k_{cp} < k < k_c$, there are two types of steady state, the development of the uniform and the aggregation of the center where possible.

Tabuchi (1986) studied the effect of the negative exponential function on the spatial interaction of a two-place system. The results showed that it was similar to the foregoing.

2. **Growth**

If the space is already in existence, the local growth is very important in the system. This polarization unit, equivalent to a seed, is a reinterpretation (3.1.12)-type of P as the aggregation intensity or density. In this way, we take the spatial dynamics process of continuous background to meet the reaction diffusion equation (Wang 1995a).

$$\frac{\partial}{\partial t}p(t,x)+\frac{\partial}{\partial x}[a(x,t)p(t,x)]-\frac{1}{2}\frac{\partial^2}{\partial x^2}[b(x,t)p(t,x)]=f[p(t,x)] \tag{3.1.16}$$

$f(P)$ is a function of space. We use $f=f(P,x,t)$ rather than$f=f(P)$, which means that the function of the space source is only related to the density of the population (in fact, the aggregation intensity), and that is not the time to evolve. Here (3.1.16) is called the reaction diffusion equation.

Wang Zheng (1988b, 1995) studied the case of the offset coefficient $a=0$, that is, the center of the place without migration. In this case,$b(x,t)=2\alpha^2$; space is a homogeneous region, $f(P)=\beta P$. There is only one effect of aggregation:

$$\frac{\partial p}{\partial t}=\alpha^2\Delta p+\beta p \tag{3.1.17}$$

P can be used to explain the self-subject density of space, and it can be found that the initial density of some space is 0, that is, the boundary conditions are

$$P(x,0)=0 \tag{3.1.18}$$

The system has only trivial solutions:

$$P(x,t)=0 \tag{3.1.19}$$

The space will never be gathered; the result means that there is space to grow, that the “fluctuation” is necessary.

If the space results from the “fluctuation” of a seed:

$$P(x,0)=J\delta(x) \tag{3.1.20}$$

J is intensity of spatial seed. $\delta(x)$ is the Dirac function. Solve (3.1.17) and (3.1.20) to obtain

$$P(x,t) = \frac{J}{4\pi\alpha^2 t}\exp\left(\beta t - \frac{x^2}{4\alpha^2 t}\right) \tag{3.1.21}$$

The type (3.1.21) indicates that the emergence of the seeds will make a general increase in space. The space effect is in the center of the continuous diffusion into the neighborhood. But the space growing condition is greater than the requirements, that is to say, the space within each point has a positive growth effect; if it is less than or equal to 0, space growth and the agglomeration effects are not possible. This analysis shows that in a developing local or regional space, to sustain economic growth, the growth results for the investment; on the other hand, fruit is a regional commodity relying on "blood transfusion" from external diffusion over the economic state corresponding to the amount of P (x, t) that does not exceed the center, the development degree of P (0, t).

C is the growth speed of P. We can conclude from $C = \frac{\partial P}{\partial t}$

$$C = P(x,t)\left(\beta - \frac{1}{t} + \frac{x^2}{4\alpha^2 t}\right) \tag{3.1.22}$$

When $C > 0$, it means that there is a positive growth in the region. When $C < 0$, it means that there is regional recession. When $C = 0$, it is a meaningful spatial position, dividing into the growth zone and the recession zone. The location r of $C = 0$ is determined by the following fomula:

$$r^2 = 4\alpha^2\left(t - \beta t^2\right) \tag{3.1.23}$$

(3.1.23) shows that the scope of the space grows and declines with time zone changes:

$$r = 2\alpha t^{\frac{1}{2}}(1 - \beta t)^{\frac{1}{2}} \tag{3.1.24}$$

When $x < r$, C < 0. The growth of the location is negative. P is in recession. There are two cases: when $t \to \infty$, $r \to \infty$, $\beta \le 0$; this means that the regional economy is lacking growth. Only a homogenizing effect occurs earlier because of external economic developments and the emergence of a sustained decline in the whole space. The situation of $\beta > 0$ is more complex. (1) When $t \in [0, 1/2\beta]$, the recession has expanded. From the point of a particle, an increasingly large regional recession, the situation can be called the hollow. However, with (3.1.21), we know that space is always at the top of the $x = 0$ point, which is not in the strength of the growth rate. (2) When R is the reduction, which means that the recession range is to the center back, the region has experienced a recession; the development of the border region has brought about the growth of the central region, a second growth in space. (3) When $t > 1/\beta$, R does not exist; with (3.1.22) the whole region is almost the speed $\beta P(x,t)$ of growth, the overall growth in space.

$\beta \leq 0, \beta > 0$ hollows out the space growth rate. We call the situation as less than or equal to 0 N (second hollow). $\beta > 0$ is the P (first) type of hollow. The N-type is a classical recession; P-type is a growth recession.

Wang (1995a) studied the case of F (P) = P in the two-dimensional case. The results show that the P is stable in the space distribution of the equilibrium state solution:

$$P(r,\theta) = R\left(\frac{\sqrt{\beta}}{a}r\right)(A\cos m\theta + B\sin m\theta) \quad \left(m^2 = 0, 1^2, 2^2, \ldots\right) \qquad (3.1.25)$$

$R(\rho)$ is the Bessel function, which is synthesized by some orthogonal functions. $R(\rho) = \sum_{m=0}^{\infty} J_m(\rho), J_m(\rho)$ is a curve of space oscillations: $= 0$, J0 at $P = 0$, and take the maximum. Zero in $P = 2.4$, P increases, J0 becomes negative, for about $P = 3.8$, and take the minimum, -0.403, then $P = 5.52$ appears at the second zero, a second maximum at 7.00, $P = 0.3$. The third negative period in $P = 865$ to 1.179. The third pole in the $P = 13.3$ nearly; its value is close to 0.36, compared to the first diode value of the rate of decline to a much smaller value. This is described by the J0 function, which indicates that the space is a ring structure; the first extreme value of P appears in the center, which is the first time. In addition, the area is in the area, the second time is gathered, the intensity P has been reduced to 40% of the central place, followed by second cycles, and the intensity is slightly less than the first third. And the situation continues; the local spacings are nearly the same, similar to this phenomenon, and the local representation is a ring structure, and in China's Shanghai, Su, Yang, often the area differentiation is very similar. When $m = 1$ is local secondary differentiation, the J1 maximum value does not appear in the $P = 0$ (here J1 = 0) and appears in about $a = 1.8$, the strength is about 0.58. J0 in the air wave curve.

In (3.1.25), we also saw changes with the angle theta, the space sector differentiation; this is the Losch found space structure. The Liao Fan structure results from a variety of values of m are complicated, and are actually observed for 3 Fan, 4 Fan, and 6 Fan, such as Shanghai is basically 3 Fan.

The discovery of the structure cannot simply be seen as Von Thunen Land Use Model, because it reveals that the city is around a central place in the formation of the lateral wave ups and downs of the city, in the urban area corresponding to the negative value of J0 negative economic attenuation. That is to say, the space is also the formation of a ring of urban intensive belt and the economic downturn. Ye (Danian) et al. (2001) first found this ring phenomenon. In fact, Ye also found some other large-scale symmetrical distributions of the city group, in addition to the ring of the center of symmetry, but also the existence of symmetrical, symmetrical, and symmetrical urban functions of the complementary symmetry of the "color symmetry" (or anti-symmetry). They further believe that in China: (1) in a city (including large, medium, and small cities) the average number of space coordinates is 6; (2) the distribution of large and medium cities in China is a lattice; the lattice of the two directions are parallel to the two main directions of the earth structure. This lattice gives evidence of the influence of the structure of the space. Ye is still in his later works to apply his observations of many parts of Russia and other places in the world to verify the universality of the symmetry of the space structure. Currently, in addition to the hexagonal structure (coordination number 6) and cyclic structure is a book mentioning the Beckmann and Puu (1990) theory, and in Wang (1995a) the kinetic theory of Shanghai Ming. Ye found a variety of symmetrical phenomena still lack rigorous mathematical proof. We say that the spatial distribution of the city has many kinds of "symmetry." The proof of the "leaf phenomenon" and universal verification are a difficult problem in theoretical geography.

3.1.3 Industrial Cluster

1. **Cluster phenomenon**

A typical phenomenon of spatial aggregation in recent years is known as a cluster. Malmberg et al. (1996) proposed a spatial economic fact: enterprises in some special industries, or in some extent interconnected firms, tend to colocate in a region, thereby forming a space group phenomenon, and group together in indemnity that does not fade for a long time. This phenomenon is known as industrial clusters (industrial clustering). Clusters of enterprises are often the production units of each industry; they have a similar relationship with the ecological community and the material supply chain, and knowledge spillovers are linked to these enterprises. The enterprises in the cluster always try to be close to the efforts of other similar enterprises and related industries, which are common economic phenomena in the real world. Further, Wang and Mao et al. (2005) believe that there are three typical facts in the industry.

1. The first Cambridge phenomenon: that is, the modern industry, especially the high-tech industry, often choses to gather, and the location of the gathering is near the University Park and other knowledge sources;
2. Cambridge phenomenon: the development of a certain size of the development of a certain scale, the development of a certain degree of stagnation;
3. Industrial aggregation is not just anywhere: it has a certain location.

Early explanation of the agglomeration economy mainly has the location theory, the external economic theory, and the economic competition process model, etc. Subsequently derived are the development of a new economic geography theory and the related mechanism research model, the competitive advantage and network cluster model, knowledge spillover and the industrial cluster model, etc.

Aggregation is the basis of clustering. On aggregation, Hotelling proposed the famous Hotelling process model. Hotelling considered the process of enterprise space under the condition of competition; the explanation of this process is seen in (3.1.3). So long as there is a fatal weakness in the process of the Hotelling process, the aggregation of the enterprise is not stable, because only one of them has a slightly lower price. It will completely exclude another firm, so it can occupy the whole market. But the phenomenon of clustering is actually there. In a new theory on clustering, to this end, Krugman (1991b) examined the effect of increasing returns on the role of the manufacturing geographical clustering mechanism. His new economic geography model is as follows: suppose that production has two elements, the nonmoving agricultural population and the industrial workers. The industrial sector is a monopolistic competition, which has two characteristics: the size of the enterprise and the transportation cost. The Krugman model is a good reflection of the size of the business expansion, but it does not explain why the business experience is in the form of independent aggregation. Peter (1997), in the development of a cluster of interpretive theory, believes that the cluster can improve productivity, can provide the driving force for continuous reform, promote innovation, and can promote the birth of new businesses. Competition is an important source of competitive advantage for enterprises. Potter believes that the cluster has brought competition, the competition of the same industry enterprises. Good performance can be obtained from the success of the honor, and poor performance or even mediocre business will feel the pressure. Continuous comparison has produced a constant incentive. Obviously, the cluster theory is the lack of a theoretical explanation.

2. **Mechanism model of industrial cluster based on Hotelling process**

Wang Zheng and Yan Xin Zhu Mao Kejing (2008) proposed a key based on the Hotelling process of industrial aggregation mechanism model; the key is introducing the spillover mechanisms of cluster. The Wang et al. (2010d) model modified the model based on the former, and summarized the general characteristics of the cluster.

Wang et al. (2010d) assume that there exists a one-dimensional linear space, which constitutes a linear market. Potential market space is continuous distribution. Technological progress is advantageous to the enterprise to increase productivity, thus the enterprise profit. The mutual proximity between enterprises is conducive to knowledge spillovers, so as to improve the technical level of a single enterprise. The toll factor is important in the location choice although it declines, but here we still recognize its influence, and as the basis of the existence of the market area. According to the classic formula (Lösch 1954):

$$r(x, d) = p(x) - c(x) - f(d) \tag{3.1.26}$$

x is the distance between any two enterprises, d is the distance of the enterprise to market, r is the profit of the enterprise unit product, p is price per unit product, c is production unit product cost, and f is freight for unit product.

First, we look at the price function. The process of enterprise space competition will undoubtedly affect the price of the product; in fact, when the enterprises are close to each other, the price of the product is the result of the interaction between enterprises, with the price of each enterprise making an adjustment. That is, the price index is a function of the number of enterprises. Here we define:

$$p = p(n) = Pg(n) \quad 1 \geq g(n) \geq 0 \tag{3.1.27}$$

p is price per unit product, n is total number of enterprises in a relatively concentrated area, $g(n)$ is the influence of factors of the price of space enterprise, and P is product prices in the course of the original Hotelling.

For the price function, the price is influenced by the competition of enterprises: the greater the number of enterprises, the lower the price. That is:

$$p = p(n) = Pg(n) \ \ 0 \leq g(n) \leq 1 \tag{3.1.28}$$

p is price per unit product, n is total number of enterprises in a relatively concentrated area, and P is product prices in the course of the original Hotelling. The significance of this relationship is that the price of the product is a function of the number of clusters and the price of the non-cluster products. $g(n)$ is the function of enterprise number. Van, Order, and Capozza (1989) stated that in an enterprise under the condition of competition, the price of the conjugate model is put forward. They define inter-firm price interaction factors for $\pi \equiv \frac{dp'}{dp}$. Wang et al. (2010d) give

$$p(n) = \frac{P}{m} + \frac{m-1}{m} \cdot \frac{P}{\ln(n) + 1} \tag{3.1.29}$$

where $\frac{P}{m}$ is the price of an enterprise in full competition,when the number of enterprises is infinite.

To discuss this theory, we assume an ideal situation, in the large-scale observation located in a region of the same distance between enterprises. Therefore, we know that in the total area, the number of enterprises can meet the following relationships:

$$n \sim (\sigma/x + 1)^2 \tag{3.1.30}$$

σ is size of space. When $x \to \infty$, $n = 1$. Companies do not gather phenomena. Now (3.1.30), substituting into (3.1.29):

$$p(x) = \frac{P}{m} + \frac{m-1}{m} \frac{P}{2\ln\left(\frac{\sigma}{x} + 1\right) + 1} \tag{3.1.31}$$

Observation shows that the effect of knowledge spillovers can reduce the variable cost of the industry. Therefore, the cost is a function of the intensity of knowledge spillovers: with the increase of the intensity of knowledge spillovers, the cost of enterprises is reduced. In this chapter, it is assumed that the impact of knowledge spillovers on cost is counterproportional, so there is the following function:

$$k(x) = e^{-\lambda x} \tag{3.1.32}$$

λ is knowledge spillovers. The type (3.1.32) of substitution is type (3.1.33). We can obtain

$$c(x) = C/\left(De^{-\lambda x} + 1\right) \tag{3.1.33}$$

At the same time, set the distance between the freight and the enterprise to the market as proportional to the distance. Namely,

$$f(d) = td \tag{3.1.34}$$

t is unit product unit distance of freight. We substitute (3.1.31), (3.1.33), and (3.1.34) to (3.1.29) to obtain

$$r(x,d) = \frac{P}{m} + \frac{m-1}{m} \frac{P}{2\ln\left(\frac{\sigma}{x}+1\right)+1} - \frac{C}{D\exp(-\lambda x)+1} - td \tag{3.1.35}$$

(3.1.35) is the relationship between the distance of the enterprise and the profit of the enterprise unit. It also reflects the relationship between the degree of aggregation of enterprises and profits.

3. **Numerical simulation and discussion**

Different types of enterprises have different costs. Therefore, the different cost of the enterprise profit with the distance change it can be obtained from (3.1.35). We might as well set $P = 1, m = 3, \sigma = 10, \lambda = 0.5, D = 2$. $C =$ 0.1–0.9. Observe the changes of corporate profits. Accordingly, we can obtain Fig. 3.1: it is not difficult to see from the picture that when$C < 0.2$, the profit of the enterprise increases with the increase of the distance between the enterprises. That is, when the enterprise profit space is relatively large, companies usually do not gather. When $0.3 \leq C \leq 0.5$, there is an optimal distance between enterprises, which makes the enterprise profit greatest, so that the enterprise can achieve profitability through a certain aggregation. When $C \geq 0.6$, not only the existence of an optimal distance, but also, if the enterprise is more dispersed, there will be a loss. That is, if the enterprise does not spill over through mutual knowledge to reduce costs, the company's revenue will be insufficient to make up for production costs and freight.

Because the enterprise's pursuit is profit maximization, the enterprise will be in accordance with the optimal distance distribution. To further analyze the relationship between the optimal aggregation distance and the cost of the enterprise, we need to solve the maximum value of the profit function $r(x, d)$. Directly use the first-order conditions to obtain

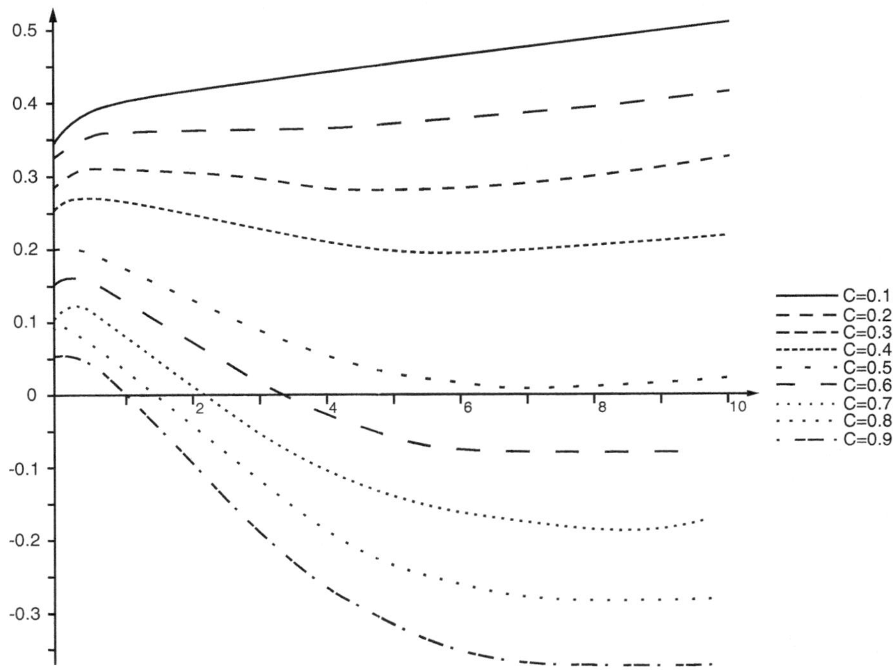

Fig. 3.1 Enterprise profit in different cost circumstances with distance change chart

$$y(x) = \frac{\partial r}{\partial x} = \frac{-CD\lambda e^{\lambda x}}{(D + e^{\lambda x})^2} + \frac{2(m-1)\sigma}{m(2\ln(\sigma/x+1)+2)^2(\sigma x + x^2)} = 0 \qquad (3.1.36)$$

Obviously, (3.1.36) type is a transcendental equation. We cannot get its analytical solution. However, we can get the numerical solution of the situation. Set $p=1, m=3, \sigma=10, \lambda=0.5, D=2$, and calculate the enterprise's profit situation, when the cost changes, as in Table 3.1. For the sake of intuition, see Fig. 3.1.

As can be seen from Fig. 3.2, the higher the enterprise cost (cost price ratio), the more it is prone to aggregation. At the same time, there is also a critical point, $C^* \approx 0.21$. Only when $C^* \approx 0.21$ will the enterprise show the phenomenon of aggregation.

To obtain the relationship between the damping coefficient of knowledge spillover and the optimal interval of the enterprise, we also use the numerical method to solve the problem $P=1, m=3, \sigma=10, c=0.5, D=2$ (Fig. 3.3).

Namely, the optimal distance between the enterprise and between the damping coefficient is an inverse relationship, $x^* \sim \frac{1}{\lambda}$. At the same time, the graph also reflects the existence of a critical point λ^*. Only when $\lambda > \lambda^*$ will enterprises have the phenomenon of aggregation occurring. This finding shows that when it is very easy for the enterprise to realize knowledge spillovers, the enterprise will not gather but will avoid competition. The greater the damping coefficient, the smaller

Table 3.1 Optimal distance between enterprises with the change of cost

c	X^*	r	c	X^*	r
0.1	2	0.42117	0.31	0.887464	0.308416
0.15	2	0.392365	0.32	0.835738	0.304069
0.2	2	0.363559	0.33	0.788948	0.299782
0.21	2	0.357797	0.35	0.707626	0.291366
0.22	1.899073	0.352058	0.37	0.63942	0.283132
0.23	1.673667	0.346568	0.38	0.609312	0.279073
0.24	1.505883	0.341318	0.4	0.555698	0.271058
0.25	1.371569	0.336255	0.5	0.375406	0.232481
0.28	1.081876	0.321894	0.6	0.274851	0.195493
0.30	0.944968	0.318518	0.7	0.212344	0.159426
0.305	0.915433	0.315638	0.8	0.170492	0.123947
0.307	0.904065	0.309733	0.9	0.140898	8.89E-02

"*" means the optimal

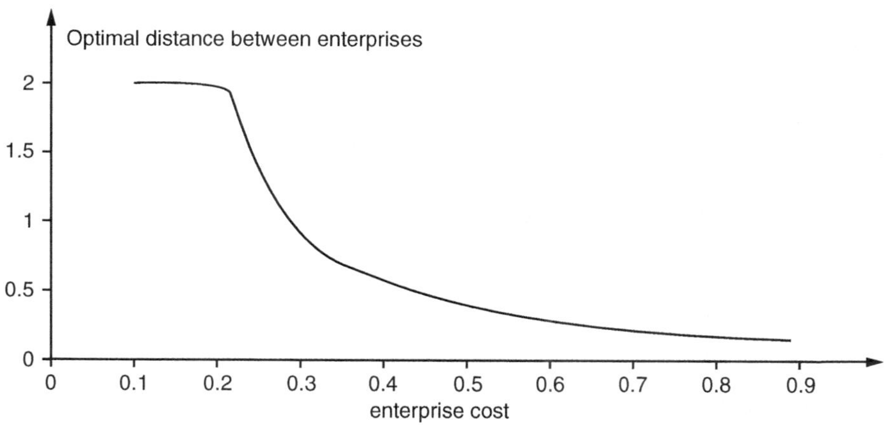

Fig. 3.2 Enterprise cost and optimal distance between enterprises

the distance between the enterprises. Generally speaking, in the early days of knowledge, the knowledge spillover is high, and it is difficult to learn. So the cluster always occurs in famous universities with strong knowledge spillovers. The first phase of high-tech industrial clusters occurs near the university, thus explaining the first Cambridge phenomenon.

To sum up the foregoing, it is easy to get such a conclusion:

$$x^* \begin{cases} \sim \dfrac{p}{c\lambda} & c > c^*, \lambda > \lambda^* \\ = \infty & c \le c^* \text{or } \lambda \le \lambda^* \end{cases} \tag{3.1.37}$$

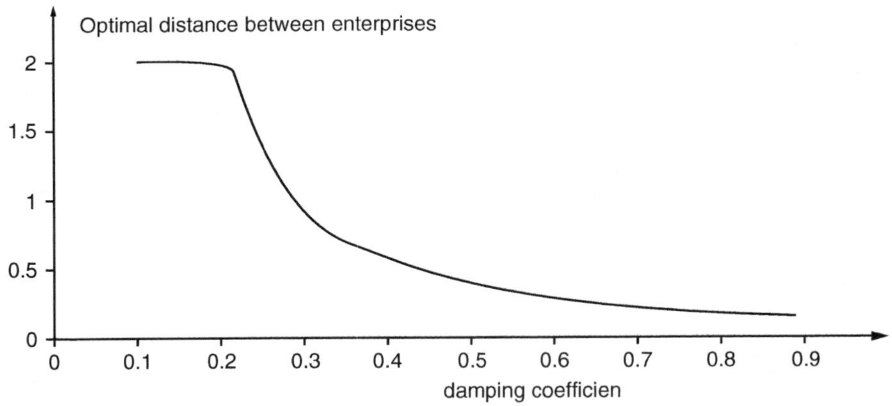

Fig. 3.3 Damping coefficient and optimal distance between enterprises

3.2 Nonnumerical Simulation

3.2.1 Cellular Automata

In Sect. 2, we discuss the dynamics problem, the mathematical basis of the reference, the physics method, and the numerical simulation of the habit; it is especially suitable for the situation of a continuous space background. In geography, however, the spatial background is often not continuous, and it is difficult to complete the analysis of classical mathematical methods. In the past few years, the rise of nonnumerical analysis is very suitable for the analysis of geographical problems. One model of the nonnumerical analysis method is the cellular automaton (CA), which has been widely used in geography.

A CA is a grid of space, each grid cell is a cell, and CA includes many identical cells (typically a thousand or more than 1 million), which are arranged in a regular grid. Meta cells can be placed in a line (one-dimensional CA), or in a rectangular matrix, even in a three-dimensional cube. In the analysis of geography, the element is used to represent a space unit or individual. The meta cell has the following properties.

1. Each cell can be in a state space to allow a state.
2. Time forward, in every step of the process, the state of each cell will change or remain unchanged.
3. The state of a cell at any time step is determined by a series of rules, which provides for the present state of how the state is dependent on the state of the cell and its surrounding cells. We use the same rule to update the status of all cells in the grid. Therefore, the model is homogeneous in the rules, unless otherwise specified to express the space of a mutation. According to the different nature of the process, different rules can be defined. For example, land use expansion and river growth rules are not the same. In the case of policy interference, the policy

can be used as a rule; for example, the city is not allowed to expand close to a range of rivers.

4. Because the rule is only a reference to the state of the surrounding neighbors of the cell, it is appropriate to simulate the interaction of the local environment in the local environment. Each cell is only a part of a conversation with a neighbor, which is directly adjacent to it.

In the realization of spatial process modeling, CA has three important concepts that need to be explained by geography: the first is the state, the second is the neighbor, and the third is the rule.

1. **State**

The state of the cellular system in a local setting may experience a set of attributes; each cell has a finite number of states, and in any time node Shangyuan cells are one of the states. For example: "1" or "0" "or "none" and so on. Generally speaking, at a time point, each cell can only have a unique state value. Of course, in the two time points before and after, the state of the cell may remain the same, which shows that the current state of the cell and its neighbor cells are not in accordance with the conditions of the state transformation rules.

Because the cellular space is not mobile, the dynamic property of the system is realized by the change of the cellular state, and the change of the local cellular state changes the global spatial pattern of the system. Figure 3.4 shows the automaton in two time points by the local state to change the formation of the global spatial pattern of change (different colors represent different states (Fig. 3.4).

At the same time, we need to pay attention to the problem of the time point of the meta cell state transition; in fact, this is also the time dimension of the problem of the automaton. We know that the automaton system is a time discrete dynamic system; in the process of running the system, all cells will undergo the whole process from time T_0 to time T_n. Then, according to the status of the update being synchronized, it can be divided into synchronous automata (cellular) and cellular automata asynchronous (ACA). In the synchronous unit cell, all cells in each of the cells at each time synchronize the status update, when the state of the system after the update of the system into the next time is T_{i+1}. In the asynchronous cellular automaton, the need is to embed a different clock in a cell, that is, different cells use different time intervals for status updates in the low-lying area with the time interval of the clock, and in the high potential area with the time interval of the clock.

2. **Neighbor**

The neighbor is a central element that will interact with another. In the (3.2.1) type and Fig. 3.5, it can be seen that the state of the neighbor cells of any element is one of the necessary conditions for the implementation of the state change rule. Therefore, to realize the dynamic of the automaton in the model, we must determine the criteria for the determination of the elements of the cellular neighbors.

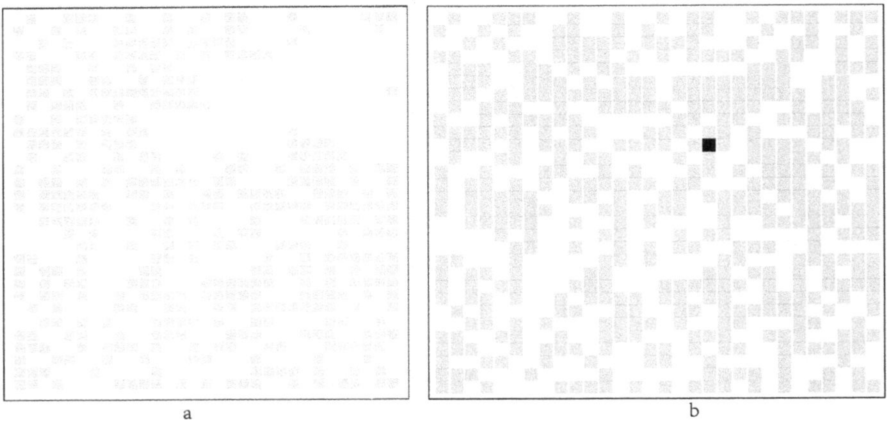

Fig. 3.4 Compare the two time-point lantern cells of the local and global state

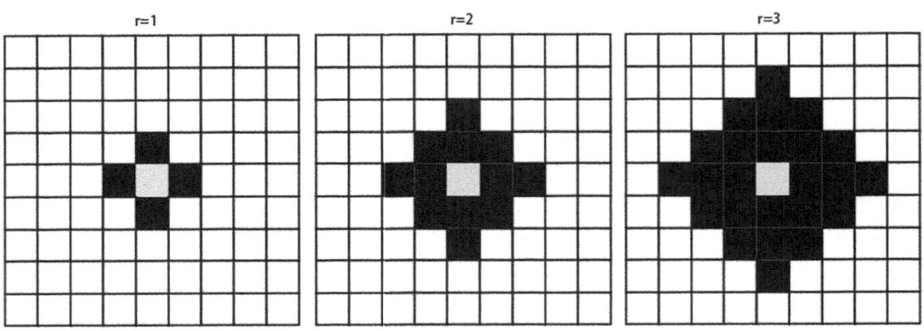

Fig. 3.5 von Neumann neighbor

In the one-dimensional cellular automaton, it is usually the case that all cells within a range of R are defined as the neighbor of the cell. In the two-dimensional cell automaton, however, the neighbor's determination to have a variety of forms is common.

1. von Neumann neighbor: when the neighbor radius is 1, the von Neumann neighbor is defined as the central element of the cell, which is defined as its neighbor; when the neighbor radius is more than 1, the von Neumann type of cell is the cell of the center of the diamond. Figure 3.5 shows:

   von Neumann type neighbor with a mathematical expression for:

$$N^V_{(x_0,y_0)} = \{(x, y) : |x - x_0| + |y - y_0| \leq r\} \tag{3.2.1}$$

x_0 , y_0 is the ranks of central elements, x , y is the ranks of other cells around the central element, and r is neighbor radius, the value for any integer. For von

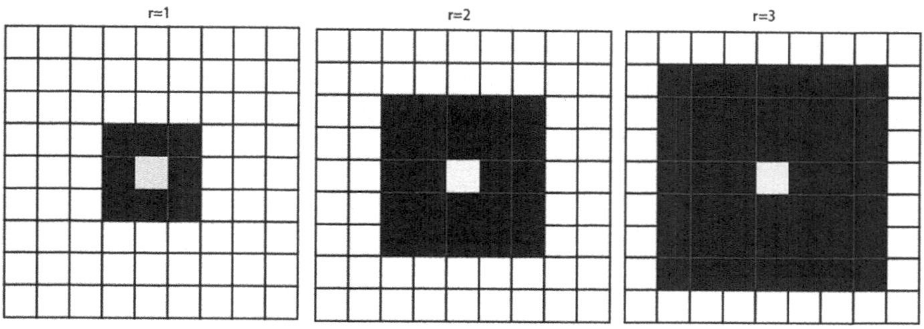

Fig. 3.6 Moore-type neighbor

Neumann, when the radius r of the cell is determined, the number of cells in the central element of the cell is $2r(r+1)$.

2. Moore-type neighbor: The Moore-type neighbor is defined as the absolute value of 8 of the absolute value of the central element cell ranks 1. The neighbor radius is 1 at this time. When the neighbor radius is greater than 1, we will be in the absolute value of the central element of the cell, which is smaller than the radius of the cells identified as neighbors, which is also known as the expansion of the Moore neighbor (Fig. 3.6). We can use the mathematical expressions of the Moore type for

$$N^{M}_{(x_0,y_0)} = \{(x,y) : |x - x_0| \leq r, |y - y_0| \leq r\} \tag{3.2.2}$$

Among the symbols and (3.2.1), the symbols have the same meaning. At this point, when the radius is r, the number of neighbor cells is $(2r+1)^2 - 1$

In short, the cellular automata are built in such a world, where the space is replaced by a balanced mesh, and as time steps forward, the world's "rule" is represented by a unified set of rules that calculate the state of its next step according to the former state and its immediate surroundings. Obviously, the nature of CA is especially suitable for the analysis of spatial processes.

3. **Rule**

The rules specify how the meta cells in the system change from one state to the next. It is the source of the dynamics of the automaton system. For specific applications, the need is for the development of different rules for the Cellular Automata, so that the evolution of the overall pattern is different. Therefore, the extraction of rules is a key step in the modeling of the cellular automaton, and is also a difficult problem based on the modeling of the cellular automaton. From the point of view of computer programming language, the rule in the cell is actually based on the condition of THEN IF. IF is to determine whether the current state of each cell and its neighbor cells are achieved by the state of the rules to do the state conversion conditions; if the result is true, then the state of the THEN occurs after the implementation of the conversion. First of all, we set the city as a CA, a CA of a local cell if there is already a known business. No enterprise is called no. Then, define the rules:

IF The cellular state is free and there are 2 to 4 neighbor enterprises in the Moore neighborhood, THEN the state of the central cellularchanges into "existing", which is defined as the aggregation of enterprises.

IF The cellular state is "existing" and there are less than 2 or more than 4 enterprises in the Moore neighborhood, THEN the state of the central cellular changes into "die"; the rule is to avoid being isolated or excessively crowded.

IF The cellular state is "die" and there are 2 to 4 enterprises in the Moore neighborhood, THEN the state of the central cellular remains "die".

The foregoing is of course a simple rule for the development of an enterprise cluster. More complicated, it can be just about a free movement to find a settlement in the space movement. Space has at least one of the settlers, which is located in the location of the cell for the " ". Now, when the migration is in, the random walk in the space CA follows these rules:

- The migrants cannot enter the cell as "cellular"
- Magix migration into a cellular, if in his field is a cellular "have," he had settled, he was the host cellular transformation for " "; otherwise, continue to walk;
- Migration: previously walk once settled, space in the "no" position randomly generated a migration, continue to walk the new walk until settled.

This rule, which is known as the space-constrained diffusion model, describes the development of the city. This simulation generates a circular fractal dimension of the urban shape. Simulations also prove that there is no first settlers, that is, there is a reason for the first to have a cell for the "there," the space is always blank, not to live, nor will the development of the city. The feature of this model is the combination of the elements of the cellular automaton and the modeling based on the self-body. Wu et al. (2011) use the CA and the subject of thinking for the simulation of climate change drivers, China's historical period, the population distribution changes, and analysis of China's famous geographical boundaries: the Hu Huanyong line appeared in the 1240s.

3.2.2 For Urban Evolution Simulation of Cellular Automata

The model and behavior of complex geographical phenomena and the self-organization system have been simulated by using simple rules to simulate the complex geographical phenomena and behaviors, which constitute the unique competitiveness and attraction of the cellular automaton. As a result, the cellular automaton is composed of a group of discrete elements; it constitutes a two-dimensional plane space, which is consistent with the characterization of the space, so it also makes research of the evolution of the city of automaton in the city deeply and widely used.

In the face of the urban evolution of the Cellular Automata, we can go through the geographical information system to obtain the geographical data and the elements of the automaton superposition, the geographical information spatial attributes assigned to each cell in the cell of the automaton. With the development and evolution of the city, the state of the cell will change. The urban evolution simulation meta cell has four possible states: (1) binary values (urban or nonurban); (2) characterization of the land using types of qualitative value; (3) quantitative characterization of spatial attribute values, such as population density, the development of building numbers; and (4) multiple attribute vector.

Because of the background of the cellular automaton, it provides simulation of the complex geographical phenomenon of the cellular automaton simulation. Therefore, the urban evolution model of the Yuan automaton needs to take full account of the special nature of the city and the original elements of the automaton to do a certain transformation. Sante et al. (2010) proposed eight aspects of the need to improve its application in the evolution of the city.

1. Irregular cellular space. The normal CA assumes that a cellular space is represented by a square grid, and some authors propose to characterize this with a hexagonal grid. In addition, the space of the cell can be formed by a three-dimensional matrix, which can be used to characterize the growth of an urban area. The rules of the grid space can be transformed into irregular patches, such as the Voronoi polygon. Irregular space units may be more realistic to reflect the objects being modeled. For example, the land parcel alternative rule cellular is closer to a realistic characterization (Stevens and Dragicevic 2007), but using the parcel unit may complicate the definition of the neighborhood.

   White and Engelen (1997) proposed to center the cell as the center of a circle, and a certain radius of cell was identified as the center of cellular neighbors. Benenson and Torrens (2004) considered that the communication between the nodes and the elements of the node based on the artificial network in geography can be used as a method for the determination of the neighbors. In the outer space, when the space division is not a rectangle, if the space is divided by the Voronoi graph, the neighbor's neighbor can be used as a criterion to judge the neighbor (Fig. 3.7).
2. Heterogeneous cellular space. In the standard CA, the cellular space is homogeneous; for example, all cells are exactly the same, but only by a different state. However, the land use types mainly depend on other land attributes, such as slope, altitude, and accessibility. Therefore, the cellular space is not homogeneous; some land may be more suitable for some other type of use.
3. Extended neighborhood. In the strict CA, the neighborhood of each cell must be the same, and it must be formed by a set of elements (such as the neighborhood of the Moore neighborhood or the von Neumann neighborhood). In urban

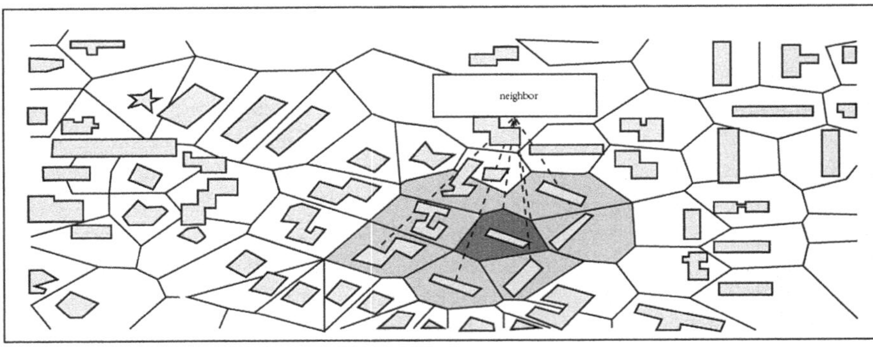

Fig. 3.7 Voronoi neighbor (Torrens 2004)

systems, it is necessary to expand the local neighborhood to consider the effects of a certain distance. When the neighborhood is expanded, we can incorporate the distance decay effect in the model; that is, with the increase of the distance between the two elements, the influence will be reduced. In the space of irregular units, a neighborhood can be defined as an adjacent unit, for example, in a certain distance of the unit or the use of a Voronoi space model.

4. Nonstatic neighborhood. The definition of the neighborhood space of each cell is different. Although this approach has been widely recognized, it has seldom been applied. In the model according to the practice of cellular state and the position setting of weight, allowing the weights to be set to zero and the introduction of different sizes and shapes of the neighborhood.
5. More complex conversion rules. The transformation rules of the normal CA only consider the current state of the cell itself and its neighbors. However, in the process of urban evolution, many factors affect the development of the city, such as sustainability of land use, accessibility, social and economic conditions or urban planning, etc. Therefore, the city of the automaton can be considered with more external factors, rather than a system such as the definition of the form of the language of the automaton. We can design the transformation rules of the city unit of automaton in different ways, and reflect different urban theories such as the theory of microeconomic planning, the central theory, or the potential model.
6. Nonstatic transfer rule. The transformation rules in the strict CA model are static, but the change of urban land use may vary with time and space. Therefore, we need to use different times and spaces to meet different characteristics. Spatial and temporal variability can be achieved by calibration. In the SLEUTH model, the transformation rules are modified according to the structure and the external parameters at each time node.
7. Growth constraint. In the traditional CA, the number of cells is changed from the transformation rules. However, urban land demand is usually determined by the

external social, economic, or environmental constraints, such as the constraints of urban growth in population evolution or urban planning.

8. Irregular time step. In many cities in the automaton, different cells need to comply with different time steps. One of the more general changes is to simulate the differences of the specific events at different time steps. Rizzi Cecchini suggests that two types of rules are applied: the application of structural rules in each iteration, and only when an event occurs.

Chapter 4
Regional Dynamics

Abstract In the fourth chapter, the author discusses the dynamics of the region as an individual, including the geography, the metaphysics of the economy, and the spatial orientation, rather than merely as a spatial phenomenon. The author believes that this is a discipline of regional science. Here, the focus is on understanding the development and evolution of the region.

In this chapter, much of the understanding of regional dynamics is derived from China's experience. China as a vast country, wherein many regional growth phenomena have been observed, but unfortunately with a lack of analytical explanation of the three natures and thee evolution of the region. In this chapter, the author's main contribution is to develop Krugman's two natures theory for three **geography**-natures theory. These geographical qualities, along with the emergence of the industrial revolution, brought about regional evolution.

4.1 Classical Theory of Regional Development

The theoretical single regional development in geography contains a variety of theories, including the theory of the man–land relationship, the evolutionary point of view of the economic geography theory of comparative advantage, and the Krugman (1993a, b) model to build a secondary nature. Space limitations prevent further discussion of these theories.

4.1.1 The Theory of the Regional System of the Man–Land Relationship

The man–Earth relationship theory, the traditional theory of geography, was originally adopted by Ratzel, Friedrich (1844–1904). Unfortunately, the first man–land relationship theory is erroneous and contains many imaginary theories, such as the effects of cold, suggesting the Russian environment results in a cold and cruel Russian. However, this thinking includes the development of environmental thought and guidance, restricting the role of culture as scientific. Later, the theory of development through proposed a "person to the relevant" theory. Chinese

Z. Wang, *Principles of Regional Science*, New Frontiers in Regional Science: Asian Perspectives 15, DOI 10.1007/978-981-10-5367-2_4

scholars Li Xu Dan and Ren Mei'e translated the relationship "land and man "to" human and geographical environment, and as "3-utline and 6-focus" stage as "three bonds six eye," the theory that human interaction with the geographical environment or the first outline of the first foucs is a human artificial construction on the ground. As can be seen, in the theory of "3-utline and 6-focus" gathered to observe the human space, development of the primary industries of agriculture, animal husbandry, and mining is discussed by different economists and geographers to evaluate the destruction of the environment by such development. However, man–land relationship students have not recognized the role of the rise and technical progress of secondary and tertiary industries.

With the increasingly prominent environmental issues, sustainable development as representative of the modern concept of this relationship has thoroughly established a dominant position in geography. Wu (1981) proposed the concept of a man–land areal system, that is, that the nature of the relationship is constituted in a system of geographical environment and human activities between the people and the land. Wang Zheng et al. (1991) proposed a "PRED" (population, resources, environment, and development) coordination view, which comprises the man–land relationship (PRED) with a coordinated core (Wang 1995b); in this system, this is the development and utilization of resources. Wang Zheng in fact uses some of the development periods in Chinese history to illustrate his point (Romer 1998). Cai (1995) and Mao and Yu (2001) emphasized that land within a specific study of human relationships is the dynamic mechanism of the people in the system, structure, and function of the human quantitative approach to system evaluation, optimization, and control of the man–land system. How to complete this development is discussed in the following views.

1. Natural resources and environment is the basis of human and social development

Natural resources and the environment provide the basic materials and energy supply for human production and life, which meets The Man and land view of the French school theory. Environment and natural resources are two different concepts, but the two are inseparable. Resources are in the environment, when the environment to provide benefits for human development can become resources. The environment and resources could become input elements of economic development, namely, natural capital (Pearce and Turner 1990). the so-called natural capital cannot be replaced by artificial capital; the natural capital has important environmental functions (Ekins et al. 2003). This natural capital determines the initial development of the region. In the early stages of regional development, humans mostly start from the basic elements, relying on local natural resources, then develop common labor; then, labor and natural resources become the first elements of regional economic growth. The factors include mineral resources and the conditions of productive agricultural topography and even ease of development. The rational use of these resources determines their long-term advantages for the development of the human condition. Therefore, in observation, the geographical distributions of population density, agricultural output, and economic activity are important (Masters and McMillan 2001); this is the first man–land relationship to determine the foundation. Also in regional economic development, human use of natural resources is constrained by skill level in natural resource-rich areas, probably because of the lack of ability for human use in harsh environments.

Chinese Han Dynasty called "the south of the Yangtze River, moist and hot, male life is relatively short" states that with the superior climate resources in the southern region, the use of technology is not enough; the climate and the rich ecosystem production potential create a hostile environment for human survival, which explains why the first civilization originated in the humid savanna grassland areas.

2. Combination of elements to produce regional differences in regional development

Production of natural resources and the natural environment as the economic factors have a part in the regional economy. This tendency to form regional differences in the sense of economic processes causes people to produce relationships with different regional manifestations. In arid regions, water is the core connection between people, and thus human use adjacent to glaciers facilitated the establishment of the Karez irrigation system. In humid areas, the invention of the canal system, by draining swamps, built the farmland system. This natural combination of elements, rather than natural factors alone, forms geographical regional differences. Obviously, these differences are not only economic but also cultural. Therefore, man-land relationship coordination is not connecting a person with single natural elements of relationship, but with coordinates the relationship between man and nature elements combined. With the traditional Chinese philosophy of thought that is "according to local conditions to action". In the coordination process, human social and economic strength and technological conditions, plus regional differences in the level of thinking, also produce understanding of each region's varying natural environment, development, utilization, and protection capability, and to some extent of the scope and content of the relationship between people and land. Therefore, the natural environment, economy, technology, and cultural level together determine the regional differences of a man–land relationship, and this is not determined by natural factors only.

3. Impact of technological conditions on the relationship between people and land

The geographical environment offers the possibility of production and life. Human skills make this possibility a reality, with technology and regional differences in different historical stages corresponding to the relationships among people of different characteristics. The neoclassical theory of economic growth for new growth and technological innovation, Schumpeter's theory, shows that technological progress has a large role in economic growth; it also provides the industrial revolution power upgrade and strengthens the role of the people and expands the breadth and depth of the land (Peter 1997, Thomas 1956). Cai (1995) argues that science and technology are important factors in the man–land relationship system, reflecting the geographical environment of human action as an important medium for the people–land relationships. As advances in exploration technology have led to the development of mineral resources, and because resource utilization is the central link in the relationship between the people and the land, human activities are associated with the natural and socioeconomic world through the development and utilization of natural resources. With advances in information technology, the relationship between people and land will undergo new changes. Wang (1998) discusses technical progress from the aspect of the historical period of China's economic development as shown by stepwise changes in the demographic nature of the bearing capacity of the stage that brings about sustained economic growth.

4. The role of Humans and Sustainable Development

People on Earth, through rational development and utilization, can promote economic development and social progress, but excessive use of resources stresses the environment beyond its capacity, causing environmental crises; although the economy is the endogenous power of human capital and technology, some resources are affected. Exploitation of natural resources caused by environmental events often has a negative spillover (Tang and Zheng 2010). The endogenous growth theory put forth by Bresthchger (1999b) emphasizes the idea of sustainable development and natural resources, both as factors of production input but also as consumer products, so that natural resources affect both the production function and the impact of consumer utility. As manufacturing uses elements from natural resources, and physical capital is inversely proportional to the changes, increased investment in physical capital will reduce the availability of natural resources. Through this relationship, the negative externalities of capital are used in the introduction of endogenous growth models. Rational use of resources and the environment to achieve sustainable development requires coordinated development of the necessary human paths.

Criticism of the early theory of Man–Land relationship is: The Man–land relationship theory describes the mechanism of the regional system relationship, but it fails to clarify the relationship between the promotion and protection of people and development of the land.

4.1.2 Comparative Advantage and Economic Growth Theory

The comparative advantage theory originated from Smith's theory of absolute advantage or so-called absolute cost theory. The theory is that every country has its absolute conditions favorable for the production of certain goods. If countries are in accord with these absolute favorable conditions to carry out specialized production and exchange of goods, they can effectively use their resources, labor, and capital to improve production efficiency and increase material wealth and social welfare among trading nations. In fact, this is a low-level difference in the natural endowments competition. The Ricardo trade theory further determines the comparative advantage of specialization and division of the region. The underlying reason is the comparative advantage of regional differences in labor productivity, but this specialization theory is insufficient to explain the actual single regional trade patterns. Heckscher and Ohlin (in Ohlin 1933), with the endowment of Ricardo's theory, introduced a variety of factors in addition to labor capital, human capital, and natural resources, having determined that the initial endowment determines the level of regional trade. Thus, the endowment theory explains well constant returns to scale in the economic system, especially for the interpretation of trade between regions within a country. Research in 1953 led to Leontief's theory of comparative advantage challenge. In 1977 Dixit and Stiglitz developed a model to explain the endogenous economy, even under the same initial conditions, not Ricardo's comparative advantage; but if there are economies of scale, the two regions can also choose different professions to produce an absolute comparative advantage. This awareness seems to exclude the effect of natural and geographical factors. Concerning this problem, the model of rigorous empirical evidence put forth by Cheng et al. (2000) proved the foreseeable, as Leontief, and the resources and technical comparative advantage model is fully compatible, thus emphasizing the development of resources.

Porter proposed in 1998 the "national competitive advantage" theory, stating that a country has a key competitive advantage that lies in the integration of the role of four basic elements and two auxiliary elements, and that these elements involve geographical factors.

Porter (1998) believes factors affecting national competitiveness factors of production mainly refer to those which a country must have, including skilled labor and infrastructure, to compete in an industry. Porter classified such elements into basic elements (or primary elements) and advanced features in two categories. The former refers to the elements of a country initially owned or those it will be able to obtain without much expense, such as natural resources, climate, geography, and demographics, and the latter refers to the elements that must be created through long-term investments and nurtured talent, such as communications infrastructure, complex and skilled labor, research facilities, and technical expertise. Porter argued that senior elements of the competitive advantage have a more important role.

Leamer et al. (1999) described the triangular role of natural endowments in economic development and natural resources. The three vertices of the triangle represent the three factors of production: natural resources, labor, and capital (human capital and financial capital). Differences in resource endowments do not indicate production of the core–periphery model but only lead to specialization or differentiation. Factor endowments or technology differences result from differences in geography, such as a national natural geographical environment (ease of access to the sea, altitude, climate, etc.); of course, these natural resources reserves constitute the underlying determinants of comparative advantage. Jones (1981) attributed the economic development of Europe to European geography and climate action. Diamond and Ordunio (2005) stated that long-term economic development needs to consider geographical and ecological factors, that Eurasian economic development mainly depends on aggregate economy and technology diffusion, and that the role of technology diffusion occurs only in an ecological belt and along the direction of longitude–latitude. Demurger et al. (2001) stated that the worldwide distribution of wealth within a country has two geographical features: first, differences between geographical and ecological conditions, such as temperate and tropical areas; and second, differentiation as emphasized through the international trade and communication leads to the regional differences, such as coastal and inland areas. The combination of these two factors constitutes a decisive impact on the economy. Bloom and Sachs (1998) found that the main reason is not the dominant mode of economic policy and economically backward (governance) conditions in the tropical regions of sub-Saharan African countries, but rather the harsh geographical and climatic conditions and demographic factors. Hall and Jones (1999) found that, because of the distance of Europe from the equator, the European economy has also a different diffusion system, which is the engine of economic development. Gallup et al. (1999) indicated that geographical differences between countries do help to explain differences in economic development.

The natural endowments theory states that the distribution of natural resources is also uneven in promoting the occurrence of specialization and trade. Based on the 1880–1998 data of 20 manufacturing sectors in the American continents, Kim (1995) examined regional production specialization patterns and found that changes in factor endowments explain the geographical distribution of most of the US manufacturing sector; however, the explanatory power of the endowment also has a tendency to decay with time, one of its weaknesses being that of "lost trade." Kim believes this is because the flow of factors causes narrow differences in the endowment of the continents and that resource-consuming consequences arise from imported resources. At this point, we can see that this aspect of natural resources in determining the status of regional trade has declined. Next, as institutional factors transformed the integration process of regional economic specialization, the customs union as a result of trade diversion destroyed the regional comparative advantage, with imperfect competition, increasing returns to scale, and the emergence of intra-industry trade, more so on the basis of neoclassical trade theory that assumes because constant returns to scale are increasingly inconsistent with reality, the role of natural resources also seems to gradually disappear. On the other hand, technological progress may change the comparative advantage. In fact, due to the development of industrial economy on the needs of energy, the emergence of all mining and refining techniques, making the arid drought in the Middle East region, has become the world's richest region, its development situation and model of a fundamental change The As a result of the progress of navigation technology, brought Shanghai from a barren village became a trade center, and created a new economic location. In other words, the comparative advantage is dependent on technological progress.

Obviously, the natural endowments theory has not explained the phenomenon of a large number of developments. Geography requires a new theory on growth and development.

4.1.3 Growth Pole–Axis Theory Point

The growth pole theory is the regional economic development concept raised by the French economist François Perroux. Growth pole theory is essentially based on the center–hinterland structure doctrine that there is room for the center, the heart of the structure. Gathering leads to some place in the economy, or the political and cultural power in adjacent areas, that has become a center. The presence of the center represses gathering in adjacent areas, so that the region is formed adjacent to the relative backward areas and has to rely on it, and become the hinterland of the center. Innovation and the development center by diffusion and spillover led to the development of the hinterland. On a single spatail form, the central hinterland structure has Thünen structural features; Innovation and the development center by diffusion and spillover led to the development of the hinterland. On a single form, the central hinterland structure has Thünen structural features; it forms a ring structure. The form of the complex point of view, the center or the heart of the structure, may also have characteristics resembling the Taylor Chris structure and hierarchical progression. We found several industries gathering around a point forming a central industrial area outside the region, such as a raw material in the hinterland region or market area, gathering into an industrial center where there may be some of the most prominent complementary industries; thus the formation of industrial circles through which several transit center rings may pass, forming an axis of industrial development.

The center, the heart, of the structure in space can be divided into four geographical spatial areas.

1. Core areas. A source has change, such as having a high potential for metropolitan economic growth or with some kind of dependence on the political and religious power (not necessarily the highest authority) of large organizations. The economic heart has a distinct advantage in the heart of the capital, technology, and policy; the population tends to be saturated or to have been saturated in politics, culture, politics, and religious fights, so that the victory or faction has basically been won.

   Up the transfer belt: in economy, for example, for the core of the impact, a good location will continue to increase investment and the degree of intensive use of resources, increasing population migration and intensifying agriculture. In cultural terms, the belief is that a new kind of people has increased, and large numbers of people have moved into the heart of the "pilgrimage" and learning; conflict has intensified old thoughts here, a revolutionary atmosphere of terror in the most conservative people.
2. Down-conversion band. This zone has a certain historical tradition of settlement. Manifestations of economic down-conversion with the survival of the agricultural area were stagnant, or even declining, aging being present within the entire industrial structure, reducing efficiency and making it difficult or impossible to intensify production methods, with a negative population migration. Down-conversion with a performance culture is a strong counterattack by traditional forces of repression, and incidents continue to occur. Some of the new confinement is enacted in the form of law, the already loose old prohibitions being reaffirmed and extended.
3. Frontier area. It is said the resource frontier zone in the economic structure of the region has a wealth of untapped resources, and such resources are potentially of great value to regional development. This zone may be located between the upward and downward areas with a belt that can be located downward. Often the first to reach this area moves upward in space and does not form a continuous strip. In the cultural structure, traditional forces loosen the frontiers of populous areas, or occupy an important position in regional economies. A large population and much economic activity connect with the outside world, especially when the center has more contact with adjacent areas, so it is possible for change to reach

the area quickly although changes in this area will be to their own demographic and economic advantage. Obviously change in the central region is the heart of the structure constantly changing form, so that changes occur in the spatial region itself. According to the central area, the heart of the structure, the development of a growth pole theory, a nonequilibrium growth theory or doctrine, encourages backward regions to set up a center so that through the growth pole formed around this up-conversion with the leading edge area and resources promotes the economic development of backward regions.

According to China, Lu (2001) proposes that a growth center is not necessarily the point, but may be a line, such as certain transportation has contacted a few developed cities, thereby forming a "growth axis." Growth axis development is often by an individual pole that is expected to cause more rapid development of the region, so that the surface region appears to have widespread growth. This theory has been called the "Pole–Axis" theory; the "Point–Axis" concept also forms a regional structure, from the point series into the linear growth incentive zone to form a linear region "center," as elsewhere in the hinterland. The point–axis theory disrupted the planned establishment of the regional production complex economic model, namely, the achievement of the country in the governance role of space economy. Letting the market have a role in the transition economies of the region is a growth theory, the infrastructure on which the local inclusive hub–network structure rests.

4.2 Regional Evolution

4.2.1 Three Natures and Evolution of the Region

From the geographical point of view, economic growth has some stylized aspects: with regional differences in economic growth in the presence of a de facto regional economy, an economic convergence, or an industrial spatial clustering, occurs. In the analysis of urban development problems, the Krugman (1993a, b) model was constructed as secondary in nature. Liu and Wang (2009) developed the idea Krugman put forward, that is, economic growth of the three geographical natures. Liu Qingchun and Wang Zheng believe the famous factor endowments that presented opportunities for economic development in the early geographical comparative advantage of the region will provide development; the persons-to-persons relationship theory is proposed to provide the basis of material conditions and therefore is the basis of physical geography. However, Krugman believes that economic development will be a self-reinforcing and endogenous factor disposition, which is second nature; they gather factors that include the region and by the accumulation of secondary traffic and location.

On this issue, we can know that second nature is a geographical phenomenon of which the most important feature is spatial aggregation, because economic development is inseparable from its support industry. Industry by enterprises with access to more profit is the ultimate goal; to reduce transport costs, enterprises need convenient transportation. Therefore, enterprises or business gatherings took place in the development of the city as a condition of their ties and facilitated regional economic development. With these conditions, cities can strengthen regional development by attracting capital and human resources, so traffic has become the second geographical nature of regional economic development. After traffic development, when aggregate market and traffic conditions occur, regional differentiation of geography is involved again, thus creating an economic location. The economic position is suitable for the development of a different industrial location; this location has guided enterprise geographically for the development of different products. Here location includes the first nature of natural endowments, but the geographical advantages for business facilitation generated the second nature, which we call a spatial position under normal circumstances, a location in this narrow sense. The location in space manifested a certain independence that can be considered an essential feature of second nature. There is also a way to talk about the relationship between geography and space. In the course of geography, we need to concur in certain things geospatially, but we emphasize that geography is spatial economics, is actually outstanding in significance in the expanded second nature. The geographical location is said to include the features of the first nature and second nature; and when discussing spatial position, their geographical background is just second nature.

Corresponding to Krugman's second nature as endogenous, the new study found that economics, the endogenous driving force of economic development, is an important factor as well as the level of human capital and technological progress, new economic growth in that they have increasing marginal returns. These factors have their regional distribution, whereby they are the third geographical nature of economic development. In economic growth, in which in turn nature actually appears in the economic process, it has at the same time been defined geographically, the so-called three natures. About a third of the nature of the earliest land avenue (2003) is clearly recognized. As Lu Avenue (2005) in the information economy, there is a new factor in the region. Joyce Teng et al. (2006) and Wu et al. (2008) presented information on the condition of facilities with especially significant impact on regional economic development, and these are independent of the natural endowments and location. Lv and Wang (2008), and Xia Haibin and Wang Zheng (2010), further proposed regional summary information facilities. At the level of the information industry development and regional human capital storage conditions, regional development constitutes a third nature: the third element of nature in regional development, together with other elements of the starting evolution of the area.

The so-called evolution of the area includes different evolutionary irreversibility plays and the historical inheritance of evolution. In a region in the process of evolution, evolution in organisms is determined by their genetic condition: the three natures of geography are "genetic." In the course of the evolution of geography, these natures might mutate, such as construction of a new transit; this gene mutation, change, and evolution will bring about changes in the direction of the regional structures. The first nature, probably the most difficult variation of the gene, so that the occurrence of certain structural characteristics remain stable, is called the lock. Analysis of weather and other features Xia and Wang (2010) found Hu's line was a basic lock of the geographical structure of China. On the other hand, global warming could lead to strengthening of the southeast monsoon, with more water vapor in western China, bringing Chinese Han and Tang Dynasty Western development opportunities; this process is called "Unlock." The first results from the nature of regional development, constraints, and other regional developments can only be second nature along constraints given the continent's development, that is to say, the development of "path dependence." Gene mutation or nature, locking and unlocking the evolutionary path of dependence, constitutes a basic conceptual system of economic geography. The study found that drive is the geographic variation of the most active factor in the nature of technological innovation and cultural innovation.

Xia and Wang (2010) studied the evolutionary economic geography point of view on regional development in China; they found that as China's first nature the geographic features of the natural endowments determine the structure of China's economic geography lock, the Hu Yong Hwan line, the Qinling–Huaihe line, and the Great Wall line, dividing the basic pattern of Chinese economic geography, cultural geography; with the second generation of the first nature on the basis of cities and roads, to guide the population, industry gathering, and regional differences in the creation of space industry. Going on to the continuous improvement of the role of regional economic development, the original lock was broken; the center city locks the basic spatial pattern, then regional centers in the hinterland of the structure emerge. Third nature emerged to promote regional structural variation; the hub–network architecture is a creation of geographical structure-mutated China. Wang et al.(2014a, b) simulated this structure. They found that, in the third nature of participation, technology diffusion is no longer the center–hinterland diffusion model, but a model of outward diffusion from the hub point along the space network. Similarly, the evolution of the spatial pattern of capital flow is no longer a standard Du Nenghuan expansion independently, but a capital hub along the main road construction of the interregional capital flow network, again through the network node to the common node of surrounding radiation. High levels of traffic arteries can promote communication between regions along the route, which has an important influence on economic development. A certain preferential tax rate will help to attract labor flow and improve the production technology level. They also made a forecast of regional differences in the future development of China and found that there is a convergence of regional development. Xia and Wang (2012) also found that in the three natures of the path dependence in space, the second nature is often the first to promote the first nature, and the third is often the first by the first nature and the promotion of second nature. On the macro level, in China's three industrial evolution processes, the Hu Huanyong line has not been broken; China's regional development has become a basic lock. In short, the three natures of geography have a decisive role in the evolution of the region. In the previous discussion, we can see the three natures of the region in the development of its geographical conditions, and that these genes are likely to promote the growth of new geographic genes, which also brings the phenomenon of the late development of the lock: for example, the natural endowment of the first nature, the city, and the development of the city to promote the survival of the aggregation and location.

4.2.2 R *Phenomenon in Regional Evolution: Locking and Variation*

The R phenomenon in the evolution of the man–land relationship is a phenomenon of regional evolution and development in China during the period of economic development in 1998, and the basic fact is that the area is an eco-economic system, with its development limited by the resource carrying capacity. Natural environmental changes, human geographic engineering, and technological advances are likely to change the resource carrying capacity of R, that is, the variation of geographic genes. R can be used to measure the population of the region, when the region's population and carrying capacity are different in different proportions, with the need of different economic policies.

Wang (1998) argues that regional growth of population is an essential factor; the regional development needs an appropriate population. For the first-time argument, geography can be from the geographical environment of the population-bearing capacity of understanding: this is actually the first nature of Shangdi's decision. The population is a biological population and cannot jump over the restriction of resources. In ancient times, technology was slow, and the regional economy was self-sufficient; the population problem is actually a problem of resources restriction.

As is well known, the population process of resource limitation satisfies the following logistic equation:

$$\frac{dn}{dt} = rn(1 - n/R) \tag{4.2.1}$$

where n is the number of population, r is the intrinsic rate of population increase, R is saturated population that maximum resources can support in population number, in theory, and n is the maximum number. The actual observation found that the population process may be the final around n for small-amplitude fluctuations of the curve, which can be explained simply as a result of random stem, which means that some point the number of individuals may be slightly greater than R. It has been proved (7.1.1) that there is only one stable equilibrium point. The intrinsic growth rate, which is not the natural growth rate, needs to be emphasized; in Eq. (7.1) the natural growth rates of demographers discussed here show that natural growth rate is a function of n. When n is greater than R, the natural growth rate is negative. From the perspective of evolutionary economics, R is a function of technological progress, technological progress, improved resource utilization, and the development of new resources; R value is improving, because technological progress is uneven, with R changes in the level of economic and population levels of stage potential transition. In this chapter, we assume that the R value of a dynasty is stable, or that it is a slow variable. Thus, when the initial value condition is $n(t=0)=N_0$, the (4.2.1)-type solution can be obtained.

$$n(t) = \frac{R}{1 + \left(\frac{R}{N_0} - 1\right)\exp(-rt)} \tag{4.2.2}$$

Theoretically, if you have a historical population time series, according to some algorithm to the historical period of the population the intrinsic growth rate and resource population carrying capacity are estimated. To accomplish this estimation, Wang (1998) uses an optimization algorithm, which is obtained from the appropriate population of the resources at that time.

The population that is fitted here is a statistical population, so the saturated population of R can be called a statistical saturated population.

From Table 4.1 it is easy to see that (1) in the course of historical development, the resources of the Chinese population are slowly rising, which can be explained as the base of technological progress. A noteworthy fact is that the population of the Northern Song Dynasty is reduced, mainly the result of the Zhongyuan Dynasty territory contraction, because it compared with the Tang Dynasty, which lost Western and North China; however, this can support the population and the Tang Dynasty compared with and without highlighting gaps because in the Northern Song Dynasty China, according to Joseph Needham's research, technological progress is obvious. Second is the Ming Dynasty; it still does not have control of the western regions, but also by the impact during the Little Ice Age, a potential decrease in agricultural production. However, it also can support a population greater than previously. These data show from the middle period of the Tang Dynasty to the early Ming Dynasty, about 700 years, compared to the end of the Western Han Dynasty, to the 700 years of technological progress at a faster speed. (2) In the course of history and human population intrinsic growth rate in the fall, this is an important cultural phenomenon. We believe that in the course of history, human beings gradually reduced the fertility rate. The nonlinear change of the R value shows that is the first natural variation.

To obtain the environmental population-bearing capacity or moderate population estimates, by the results presented here, we can analyze each regime change in the war; in the population drop from stable political conditions, the population tends to be a saturation population in various stages of the duration and the characteristics of social development.

The table shows the Western Han population takes $3{,}000{,}000 \times 5 = 1500$ million. The starting time for the war period has been stopped; in fact it is not sensitive to the initial estimation calculation, the seventh column analysis conclusion.

The analysis (4.2.2) is of great benefit to our understanding of the role of R in saturated populations. First, we notice that the population of $0.8R - 0.9R$ is the age of the population (Table 4.2). It is located in the Han Wen 14 years; the emperor,

Table 4.1 China history: intrinsic population growth rate and population resource saturation

Age	Since the year	The year ended	*R* (0000)	The intrinsic rate of population growth (%)	Residual variance
Han	In 2 years, AD 57	Yongshou 3 years, AD 157	5100	6.15	50,881
Tang	Kaiyuan 14 years, AD 726	Tianbao 14, AD 755	5670	5.65	1177
Northern Song	Yuan on the first year, AD 1086	Grand 4 years, AD 1110	5220	4.95	946.72
Ming	Yongle 11 years, AD 1413	The first year of Taichang, AD 1620	6750	2.19	71,469

Table 4.2 Historical dynasties corresponding to the population characteristic value, *S* (the number of years in brackets)

Dynasty	The Western Han Dynasty	The Eastern Han Dynasty	Tang	The Northern Song Dynasty	Ming	Features
Starting time	Before 202 (0)	57 (0)	627 (0)	980 (0)	1413 (0)	Recovery phase
0.8 *R*-years	Before 165 (36)	85 (28)	669 (42)	1010 (30)	1425 (12)	
0.9 *R*-years	Before 152 (50)	99 (42)	684 (57)	1025 (45)	1462 (49)	
0.95 *R*-years	Before 140 (62)	111 (54)	697 (70)	1041 (61)	1496 (83)	Boom stage
0.99 *R*-years	Before 113 (89)	137 (80)	726 (99)	1075 (95)	1570 (157)	Peak stage
0.999 *R*-years	Before 75 (127)	175 (118)	767 (140)	1122 (142)	1675 (262)	The decline stage

5 years, Han Zhang Diyuan, and 2 years to Teijin Nagamoto for 11 years, Emperor Gaozong General. Two years to the first year of daylight house, song Zhenzong big Zong Xiang Fu, 3 years to Renzong Tiansheng, and Emperor Yongle 22 years to the Pope Tianshun 6 years. We can see the corresponding age is the full restoration of these dynasties, the economic Shengping era, in the Western Han Dynasty, which corresponds to the early glory "Wenjingzhizhi" in the Eastern Han Dynasty. It is considered to be the "old Beadon Ren, people Yanfu, cattle and sheep everywhere" of Sheng Ping times. In the Tang Dynasty the Empire is strong and expanded to the East and West in the two directions, and in the Song Dynasty it is a blessing and contains many legends of prosperity. In the Ming Dynasty, it was a time when China was so powerful that it could be extended to the outside. We believe that this

is the main reason for the prosperity of the first generation and the second generation of the population growth in the post-war population that has grown to be the labor force. Labor supply is adequate, when the population has had a large increase in the number of people, and the community has a full and effective demand to promote the development of production. Notable, at this stage, in addition to the Ming Dynasty, is that the China orthodox advocate "restraining" Pro Confucianism has not been fully emphasized or commercial activities encouraged. At the same time, because of the distance between the saturated population of 10–20% of the gap, the large natural disasters will not cause a stage of population surplus. Actually Wang Zheng and Zhou (1995) found that even up to 1816 the serious lunar June Anhui snow climate mutation-type natural disasters only caused crop decreases by 10%.[1]

The next population of the 0.9–099R stage, generally speaking, is the stage of economic prosperity: in the Western Han Dynasty, the Han Emperor for 5 years to Emperor Yuan Ding is 4 years. "Wenjingzhizhi" continues, and the beginning of the Western Han Dynasty was the prosperous Emperor Period in the vicinity of 0.95*R*. In the Eastern Han Dynasty, the period for the Emperor and 11 years to Shun Emperor Yonghe is 21 years. During this period, the Central Plains Dynasty reoccupied the northwest area. Cui Shangde, a reckless one (1992), studied this era as an era of economic prosperity, the social emergence of an unprecedented luxury of the wind and the general consumption of luxury, commercial, and economic development.[2] In 697 years of the Tang Dynasty and 0.95*R* corresponding is the magic days after the first year, before and after this era is Wu Zetian career development to the peak of the era. Although the 0.99*R* corresponds to the 726-year Xuanzong Kaiyuan is 14 years, Kaiyuanshengshi is just the midpoint of the Tang Dynasty prosperity. In the Song Dynasty, 0.95*R* corresponds to Yu Renzong Qingli, 8 years of 0.99*R* corresponds to Shenzong Zhaoning; this is produced "riverside" to describe the age of prosperity. In the Ming Dynasty's corresponding respectively is 9 years and Takamune Hiroji Mouzon Longqing 4 years. There is no reason to deny the prosperity of this period. In particular, it is in the late period of the late 1980s, the society generally produces change, it is the last stage of reform, the event. In the Western Han Dynasty then was held the meeting of salt and iron, in the Northern Song Dynasty, Wang Anshi. In Wang Anshi at that time many modern historians accepted the capitalist measures[3] for Zhang Juzheng's financial reform in the Ming Dynasty. Because at this time the population reached or was approaching saturation, the lack of resources, the absence of technical change, the promotion of consumption, and promoting the business of non-Confucian ideas will not necessarily receive a general welcome, so

---

[1]Wang Zheng and Zhou Qingbo, et al. 1995. Nineteenth Century on the first half of the leaf of a climate change. Natural Science Progress 5(3):323–329.

[2]Cui Shangde 1992. Cambridge, China, and the Qin and Han Dynasties. China Social Sciences Press, Beijing, pp 648–656.

[3]Deng Guangming 1979. Wang Anshi. People's Press, Beijing.

it is always conservative to achieve the final profit. In the prosperity of China's early accumulation of wealth, no policy support will lose the pursuit of commercial interests and the development possibility of production and technological change; China's misfortune is how it happened.

As a matter of fact, the population of the 0.99*R* is a critical population; any small natural disasters to generate *r*, a transient decrease in all, will enable population excess to ensue. The period from 0.99*R* to 0.999*R* is bound to be a period of decline. The potential political reform, technology, and management system of innovation are expected to allow the economy continue to grow. But the changes of the Chinese dynasties have failed in the conflict. The Tang Dynasty is no reform conflict, ending with the Lushan rebellion. Dynasty was in its peak with the population boom. Is a fragile system, this time once the climate change occurs, the harvest will fail, and a large number of population will be surplus immediately.

Overpopulation contributed to the peasant uprising, the Western Han Dynasty, the Eastern Han Dynasty; Mingdu corresponds to a 0.999*R* and appeared shortly after the outbreak of the peasant war. As Wang et al. (1996) points out, the result of this war will inevitably lead to the warlords and war; the population declined shortly after a change of dynasty. It is worth noting that the 0.999*R* of the corresponding period did not appear in the Northern Song Dynasty, in other words, only from the resource supply conditions; the Northern Song Dynasty is not the end of its own, and the foreign achievements are a period of their own development. This point can also explain why the Song Dynasty king in the political fight against the Jurchen people eventually lived and continue to own the southern dynasty. In contrast, the Ming Dynasty in the attack by the people of Manchuria was soon destroyed with the population surplus and the peasant war caused by the depletion of the Ming Dynasty's national strength. Tang in the era of 0.999*R* corresponding appeared to have long perished, because the causes of population plummeted during the rebellion of the population and did not actually reach the previous level. As a kind of external interference, the war is changing the population: "the external cause is changed into internal factor," which influences the regional evolution.

From the geographical point of view, the Song Yuming is an unfortunate dynasty. After the 880 AD abrupt climate change, the Yellow River suffered from increasing,[4] the court had exhausted the state finances to harnessing, and in errors from the early Shi Jingtang the Central Plains lost the north wall barrier, the river, and anti-aggression so that countries do not have sufficient financial resources to develop production.[5] The tragedy of the Little Ice Age began in the past 500 years: China LIA experienced three waves. This kind of fluctuation makes

[4]Xu Hailiang 1993. The history of the Yellow River downstream river sediment variation and change. See button Chunghsun editor, the Yellow River water and sediment run. Environmental Change and Rule of Proceedings (d), pp 68–76. Beijing: Geological Publishing House.

[5]Lihua Rui 2006, Song River and anti-edge. Song and history research, pp 136–153. Ancient Books Publishing House, Tianjin.

the crops in the early stage to adapt to the climate of the late, not to adapt to the inevitable performance of natural disasters. Such Song and Ming Dynasties in their prosperity did not reach the Han: Tang's glory is understandable. As the *R* value is a function of the state of resources, it is a function of agricultural resources in ancient times. The fluctuation of the *R* value is the fluctuation of the climate fluctuation, which is inevitable. In addition, another reason for the failure to achieve the desired prosperity of the Ming Dynasty may be policy. Ming Dynasty began to "weigh this curb not," the policy does not encourage a commercial "inaction" stage, did not let the second industrial development and promote regional new geographic gene growth. We call this phenomenon the "*R* phenomenon"; in the economic development of resources to achieve the maximum carrying capacity of 95%, the economy appears "flourishing," then the economy is facing a threat, and then in the resource utilization to reach the maximum carrying capacity of 99%, the economy sees an almost unstoppable decline.

Several eigenvalues are obtained by observation: 0.8*R*, 0.9*R*, 0.95*R*, 0.99*R*, and 0.999*R* in turn. For some of the major dynasties in Chinese history, the population has been calculated and the recovery of the population began to be stable, and the time of the characteristics of the value of 4.2. We find that although the *R* itself is caused by technological progress, it is still in a certain rule to lock up the development trend, which is related to the characteristic values of 0.8, 0.999, 0.95, 0.99, and 0.9 of 4.2.

For these "variabilities" of the coexistence of the *R* phenomenon, Wang (1998) proposed the following economic commentary: in mainstream economics, consumer demand determines the economic growth, the limited resources are not considered; this implies that as long as there is demand, supply will always be created. This is, of course, the case for the full guarantee of resources. Sustainable development has to start from resource utilization to achieve the maximum capacity of resources. In fact, given that the resources cannot provide unlimited supply, in certain technical conditions, with certain resources, its carrying capacity is certain. Resource-bearing capacity determines the maximum value of the requirements. We analyze the economic process in Fig. 4.1: Y states the national income, and D is demand; this is Keynes's theory. First, we have a "golden age," as we have described; the use of resources has been close to its carrying capacity (where the value of the carrying capacity, in the case of no account, will cause the actual *R* decline, we take the time has reached *R*). At this time the society has reached the level of demand AD. Demand for national income is further growth, the demand for D will be further growth, and a new social demand level AD is put forward. The problem is that the AD is the highest requirement of the resources under given conditions. However, the development of anthropogenic trends continues, which includes population growth, personal consumption level, and expanding investment, resulting in the appearance of "excess demand." In a closed economy, there are no external resources, and the excess demand is not achievable: the new equilibrium will never be reached. However, the population is growing, so the results can only be a decline in per capita national income. Historical data show that with the development of technology, the *R* value of China is increasing, E is

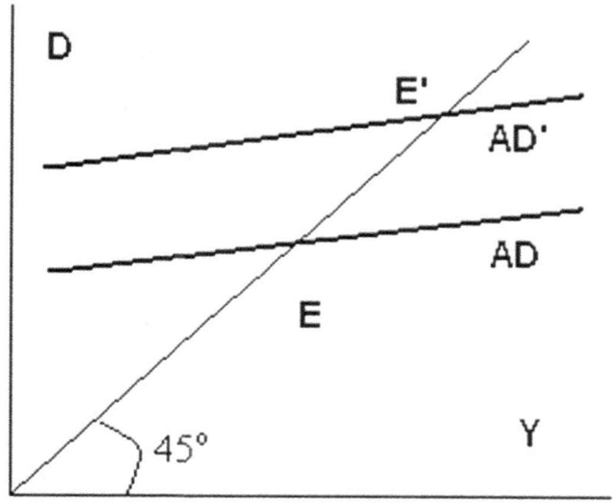

Fig. 4.1 Production of excessive demand

increasing; the Chinese economic level is constantly developing. In other words, technological progress is the fundamental guarantee for economic development to reach a new level. Keynes's theory policy and the theory of supply school are only effective in regional development.

In the economic system, the economic level is measured by money. G for the economic development level of the actual monetary aggregates, according to the Fischer equation G = MV, where M is the amount of money, V is the money flow rate. At the equilibrium point in Fig. 4.1, $G = MV$ is the total amount of money used by the society. Because supply is the need for resources to ensure that N represents the total supply of resources to support the total supply of goods, P is its price in a stable economic system, with supply and demand balance, and so

$$G = MV = pN \tag{4.2.3}$$

When resource utilization has not reached saturation value, or the economy has reached its peak, then we follow the general growth policy, increase the money in M or V, and can ensure the further development of resources and expand the production scale. Because the resources are guaranteed, and improved consumer demand can make the capital owners dare to take risks, the results G and N also increased, for economic development; this is alright but N is not up to the saturation value of resource utilization. If you do not change the price system, there is G > pN. To achieve a balance, the (4.2.3)-type equation is set up; the need to develop the resources to make N rise, the driving force of the development of such a stimulus to achieve, so the stimulus spending can protect economic growth. However, when N has reached the maximum of resources to support the development of total N_0 (R), N will not increase again. At this time prices can only be generally raised, as price increases equal currency depreciation. In other words, if we follow the general growth policy, increase the money investment or the flow rate, we cannot expand the production scale, because the carrying capacity of resources has been saturated, and increase in the amount of money just brings inflation, and we are wrong to increase unemployment. This is a typical stagflation, in the early stage, of the policy model of Keynes's doctrine, which had the opposite effect. Different stages of development require different economies, which are formed by the different limits or levels of supply. The role of resources is to lock up growth; regional development is subject to the evolution of the geographical nature of the constraints, not arbitrary.

To history as a mirror, in the "spirit of the economy," the use of resources has reached 0.95–0.99*R*; Keynes's theory on stimulating consumption can promote economic growth but the conclusion cannot be set up. The conclusion can be obtained from the economic development of the Western Han Dynasty. At the beginning of the Han Dynasty, the "Huang Lao philosophy" encouraged circulation and consumption. Ideological circles think: "Xing of how, it is good at luxury"; "don't waste cannot be established" ("pipe-luxury"). This kind of thought and Keynes are a "general consensus." Then, the Chinese economy develops quickly. By the late emperor era and not detected early in the local and national competition for resources (copper, iron, salt), the resource supply shortage is highlighted: "agricultural Chen Cangqian consumption, tax is exhausted" ("records of the historian: flat accurate book"), with local sustainable development problems. In the Han Emperor Showa era, the conference resources (Salt iron meeting, Confucian scholar in the Salt iron meeting to the state proposed "civilian luxury is hungry"), "may strike the salt and iron and Wine Monopoly, are lost, so Jinbentuimo, the agricultural economic proposition, this claim and supply experiences relaxation of controls (deregulation), tax cuts, and the idea of production development." Bringing economic prosperity of key economic officials in Sang Hongyang is proposed to rely on technological progress to develop resources, promote circulation to keep sustainable development advocated: "state-owned Woye Rao Ermin insufficient to predators and instruments unprepared also. Have the goods and people to Shanhaiguan property and not too." Salt iron conference results are a national fee except for salt production control and licensing system, but also to maintain the Sang Hongyang's other economic proposition in Sang Hongyang and from Guanzi's consumption thought backward: "gentleman luxury thorn frugality, thrift is solid." ("Yantielun").[6] The Han Emperor said: "The Chinese own system, the combination of law and Confucianism, but as pure Confucian benevolent, with Zhou Zheng?" Han Zhao, the Xuan emperor, became the master of ZTE; they found the Chinese Empire "sustainable" economic policy, that is, "miscellaneous." We know "stagflation" appeared in the 1980s at the beginning of the United States; the supply siders policy changed the economic process. One of its characteristics is lower taxes, the implementation of the economics of the relaxation of regulation. America's economy has recovered. Visible tax cuts and deregulation cope with the phenomenon of *R*, because it may be through the use of resources to improve the *R* value, so that economic growth

[6]This paper used the "pipe" for Beijing Yanshan press 1995 edition. Jilin Yantielun Literature Press 1995 edition, Wang Zhenmin.

occurs. This policy is a change in management. When the Yantie conference of scholars actually proposed tax cuts, relaxation control but not with “luxury thorn thrifty” policy at the same time, then followed by the economic recession, economic policy debate went into political retaliation, the “luxury thorn thrifty” policy under the condition of “Confucianism only” ended; the winner did not seek a new economic path, but the implementation of ethics “to protect, to ethics to control economic development requirements.” Emperor Xuan of the Han Dynasty to advocate Confucianism, Prince later Emperor, said: “chaos of my people, the Prince.” History, unfortunately, he said was the result of not expanding production through technical progress; in Confucianism, let the ideals of the intellectuals to become civil servants, economy is inevitable decline, the state corruption is inevitable, the Emperor himself had to give his beautiful wife Wang Zhaojun to the Huns. Of course, the problem has two sides: the Confucian doctrine advocated saving resources, when it later become an ethical thought; it may become one of the ethical foundations of sustainable development (Wang and Wang 1998).

4.3 Regional Development

4.3.1 Overview

The early development of the region is the symbol of regional urbanization. The city is not the ancient city, the ancient city is a resident, in the clan society; the same clan to live forms the initial military self-defense, then the city has no special economic significance. In the modern sense, the city is an economic unit, which is the product of the commodity economy. The function of “market” is the exchange, the exchange of goods; the reason why we need to exchange goods is because in the regional production of the products with their own characteristics, the occurrence of characteristic products follows the differences in resources. The trifoliate orange was born in the Huainan orange; in fact, orange and trifoliate orange are different species. Their adaptation is to the Huaihe River to the South and to the climate in the North, so that Chu will be orange and Qi will be trifoliate orange, then farming patterns also vary. Then, there is a mode of cultural differences. Because of the diversity of consumer demand, people want to swap the orange and citrange, a trade behavior. This kind of trade behavior needs a place, for each city “opening.” Economists see this as the occurrence of a specialized division of labor. Also, different environments lead to the initial division of labor and specialization and cultural pattern differences: this is the man–land relationship automatically coordinating decisions.

In the region, we often see the man–land relationship showing a coordinating role. In this story we can see that the city is “opening” to urban transformation, and opening specialization further. Because of trade events, it is the pursuit of efficiency. Considered from the point of view of production, products are less diverse, producers more easily move in a direction of technology excellence, improving efficiency; at the same time, the production of service efficiency is high, so this production will be able to expand in scale with low cost, resulting in efforts more conducive to trade benefits. One can say that the development of urbanization occurred simultaneously with the regional division of labor and specialization, that these are the cause and effect of each other. Here we also see the phenomenon of early development. Regional early development from a self-sufficient agricultural economy began the development of transformation to an industrial economy, changing the size of transactions, and here the City as a trading center is indispensable.

In the early stage of development, an important factor that cannot be ignored is the accumulation of industrial capital, and this accumulation is from the beginning of agricultural surplus. Observation of China's urban development, initially in the city focused on a group of people engaged in manufacturing, can have a specialized population engaged in manufacturing, indicating that there was a surplus of agriculture; if the total population of agriculture is not sufficient to solve the problem of food, and there is no agricultural surplus, manufacturing is difficult. We know that in the slave society, which is based on the cruel exploitation of slaves, there have been artisans, free people, and nobles; the manufacturing industry has not developed because with the essence of the agricultural surplus it is impossible to accumulate manufacturing capital and support technological innovation. Engaged in manufacturing, centralized management, and craftsmen communicating technical information to the city in the dynasty, the craftsmen products, in addition to supplying the nobility with good for consumption, can also provide other people with goods, concentrated in the city trading so as to reduce the transaction costs, so the transformation is from city to city. However, in ancient times the most developed were those trade cities. For example, in Yangzhou in the Tang Dynasty is located in the north and south canals, which is the Yangtze River Interchange; the Yellow River, Huaihe, the Yangtze River, were three river basins, and with the exchange of goods in the Taihu river basin hydrology-geomorphology hub, an important role position of Yangzhou occurred in the trade status. The geographical position in regional development has an important role in the early stages as an expression of the relationship between people and land.

From the prominent role of trade and the aggregation of the population, the development of the city is limited by the scope of the market, so the city began to expand; the expansion of market size is not only caused by population growth but also by the expansion of transport equipment and services to the regional market: the consequence of this process is the occurrence of an industrial revolution. Landes (1969) pointed out that many technical improvements have been made to understand industry, and the invention of the steam engine is a classic example of this. In the nineteenth century, it was known that the division of labor had effectively improved productivity and had begun to look for ways to improve technology. In this way, the industrial revolution began to accumulate the technology of self-sustained growth, so that the world's population grew faster than in the past, by eight fold. The first time to explore the wealth of technological efforts to open up the market, and promote industrial production, was with the development of the economy.

4.3.2 From the Primitive Economy to Aggregation and Industrialization

Goodfriend and McDermott (1995) produced an early development model. This model assumes that there are two kinds of production: one is the family as the unit of agricultural production, which produce the final consumer goods, and the other is modern economic production. They need to be in the middle of the market transactions. The existence of intermediate goods means that the region has a specialized division of labor, the market as the center of the industrial economy, the two economic sectors being the market economy and family. There are three characteristics of the original technology and the market production process area. First, each family has its own original production process, but the market is from all the families to hire workers. Second, the single family is too small to use the original technology. Third, human capital ($h = 1$) cannot increase the productivity of the original production.

In particular, it is very important to point out that there are many models of regional industrialization and urbanization, but the GM model (Goodfriend and McDermoot 1995) is based on the new economic growth theory. The key idea of this model is shown in Figs. 4.2 and 4.3.

In Fig. 4.2, we see the economic system in the presence of two kinds of economy: the use of the original production of the "family subsistence economy" and the market-oriented production of the "industrial economy." Labor to invest in the market economy sector "efforts" (labor time and total labor share) is e_M, and to invest in the original sector "efforts," e_P, the sum of 1. In Fig. 4.2, the marginal product of "hard work" at any point on the original branch of the family subsistence economy $e_M(=1 - e_P)$ is greater than that of the market sector (w). It can be proved that the mixed economy cannot be e_M, e_P more than 0, because no one can guarantee that the two departments have the same marginal utility and make the effort of the division of the department and the one so that only the C, D point can be a balanced point, but the D point is not stable. Once the individual families do not pay all efforts into the market sector, but only pay part of the original effort, they will achieve the same or even more; the family will be transferred to the original department, to maintain the original revenue that reached E point, then the market sector can have benefits at point F; it is less than the benefits of E, so these families will continue to transfer to the original department, and finally all the economy-only C points correspond to the state, although they are kept at D point.

Figure 4.2 only occurs in the pure original family subsistence economy. The conditions of regional industrial production, as shown in Fig. 4.3, are first balanced; in Fig. 4.3, the market sector wages are equal to the marginal returns of the original

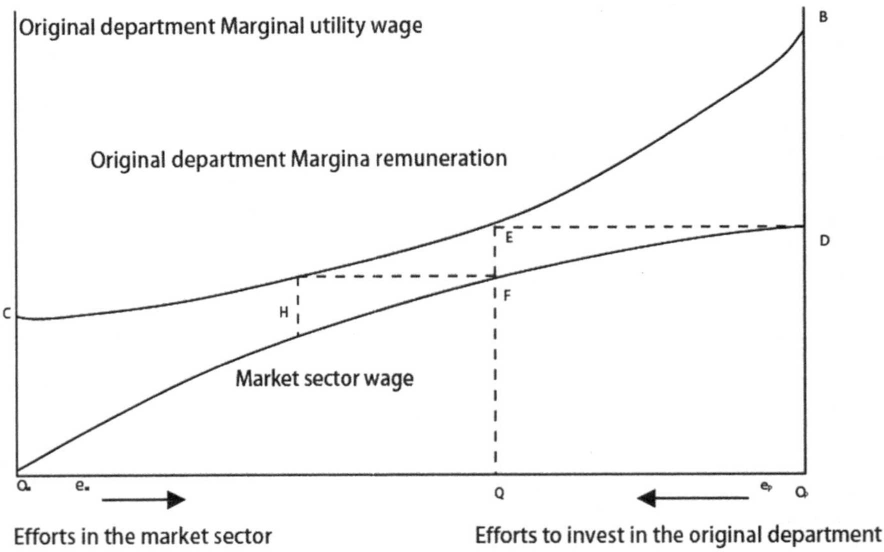

Fig. 4.2 Pure original sector economy (From "The marginal utility of the original department B. Marginal revenue of the original department; market department wages")

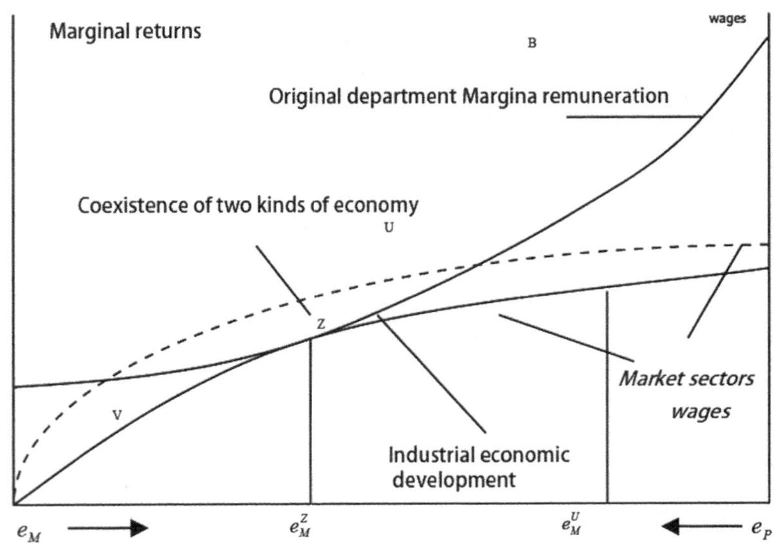

Fig. 4.3 The occurrence of industrialization

sector. Second, when other families distribute to the timing, in a balanced economy (Z) efforts to allocate (e_P,e_M) will not occur because a single family is stimulated to deviate from the economy. In the Z point of the economy, if the market sector seeks to further improve the efficiency of the broken state (or the beginning of the initial sector is decreased), there may be specialized production. In the Z point of the economy, if the market sector is to further improve the efficiency of the broken state (or the original sector decreased), professional production is possible. To improve the efficiency of the conditions, U, in the market sector, the V segment appears to be more than the original market sector wages, the occurrence of industrial economy. The GM model proves that the condition of the initial market sector benefit is open to the market, and the population size is enlarged, which is related to trade. From Fig. 4.3 are found some characteristics of agricultural civilization that may explain the development of agricultural civilization:

(a) The population must grow to a threshold to support a city's market sector, the emergence of the "city" to the city's transformation; this time the population growth has brought urbanization, the formation of the industrial economy, as the basis for industrialization. The population must reach a second threshold to start the industrial economy.
(b) At the beginning of the industrial economy, the accumulation of human capital and knowledge did not start until the market size of specialized commodities was expanded.
(c) Population growth is helpful to industrial economy, but the starting factor of industrial economic development is human capital and specialization, the aggregation of human capital; the economic growth or attenuation of the population depends on the size of the market.
(d) Initial conditions for industrialization to begin are important.

On the Goodfriend and McDermott (1995) (GM) model from the primitive economy to the industrialization of 2002 is the theory of economic geography. In the Goodfriend and McDermott (1995) model, aggregation and industrialization are at the same time.

4.3.3 Regional Specialization

In Sect. 3, we see that specialization is the basis of industrialization, which is not the only case that leads to a strong interdependence among the various regions; thus, regional integration is the process.

The regional professional sector turned into a kind of universal phenomenon that was noticed by many scholars, and a large number of theoretical models have been developed. William Petty in the seventeenth century at the earliest recognized the significance of professional productivity in the eighteenth century. Smith divided the work system and put forward the important assumption of specialized division of labor: the level of division of labor is determined by the size of the market, market size is determined by transport efficiency, and capital is an indirect production sector development division tool (Yang 1999). According to regional specialization, Smith proposed an interregional absolute interest principle: Smith thought if regional trade occurs it will choose the dominant industry to be developed, the absolute advantage may be given the structure of regional resources and environment that occurs with specialization. After Smith, Ricardo put forward the regional comparative advantage principle in which he envisaged two regions; the production of different products in different regions are not the same as to cost; there is a price ratio between the products, because of regional differences, two of the same products in different regions of the price are not the same, because of the comparison between region and region, the production of different products of interest are not the same, the price here having a comparative advantage. The existence of the regional comparative advantage also led to the development of a regional division of labor. Benedikt and Stiglitz (Dixit and Stiglitz 1977) developed an endogenous economic model to illustrate that, even in the case of the initial conditions, there is no Ricardo's comparative advantage, but the two regions can also choose different professionals and produce the absolute advantage. This book mainly discusses a typical model of the new economic growth theory, which was developed by Desmet (2000).

Desmet (2000) considered a two-regional economy, the East and West; the economy has two sectors, the food and manufacturing industries. As a result of early development, and regional division of labor and labor skills, while labor is the only factor of production, and only two kinds of technology, they are not completely replaced; the two regional population flows may be the result of changes in income and the existence of population transfer.

Desmet agreed: the variable superscript * indicates the west, variables with no asterisk not only are the eastern of the two areas. The East and the West have the same labor, and agree as a unit, that is

$$L = L^* = 1 \tag{4.3.1}$$

Because workers can move between the two places, some workers have the food industry, but may be employed in the manufacturing sector, some workers have manufacturing technology, but may also be employed in the food sector:

$$Q_F = L_F^f + L_F^m(1-\theta) \tag{4.3.2}$$

$$Q_M = A\left[L_M^m + L_M^f(1-\theta)\right] \tag{4.3.3}$$

Among them, Q_F and Q_M are the numbers of food and manufacturing products, and the skilled labor in the department; as the number of products in the food industry with the food industry, the output of the unit labor force of 1, said manufacturing in the manufacturing sector with the output of the unit labor (Eqs. (4.3.2), (4.3.3)). Workers with manufacturing and food technology are not entirely alternative; manufacturing workers have lower productivity than manufacturing workers in the manufacturing sector, with food technology workers in the manufacturing sector, as well as productivity. Western technology and Eastern technology are the same, so the West also meets the same relationship. Because the price of food is 1, the labor market is perfectly competitive:

$$w_F^f = 1, \quad w_F^m = 1 - \theta \tag{4.3.4}$$

$$w_M^m = p_M A, \quad w_M^f = p_M A(1 - \theta) \tag{4.3.5}$$

which is w_j^i employed in department j of skilled worker wages, and prices of manufactured products. With the absence of transaction costs, free trade is equal to the cross price, so workers at the same wage are equal. Given the manufacturing and food sectors are not completely replaced, people are generally based on their technical expertise to work in the relevant departments, so the relative supply of manufacturing is $(A(L^m + L^{m*}))/(L^f + L^{f*})$. However, from Eq. (4.3.5), if the price of manufactured goods falls, there is no so-called work in each department. If prices fall further $(1 - \theta)/A$, manufacturing products will disappear completely. Similarly, if the price of manufactured goods is rising as $1/(A(1 - \theta))$, the workers with food technology have no so-called work, higher than the price; so food production will also be completely stopped.

The relative demand for manufactured goods is determined by the consumer's preference, and the utility function is determined by the utility function of the Douglas type:

$$\frac{C_M + C_M^*}{C_F + C_F^*} = \frac{\mu}{1 - \mu} \frac{1}{p_M} \tag{4.3.6}$$

The equilibrium price of manufactured goods is equally obtained by the relative demand and supply, such as in Fig. 4.2, in which the relative demand curve intersects perpendicularly to the relative supply curve. If all the workers are working in their corresponding department, then $(A(L^m + L^{m*}))/(L^f + L^{f*})$; the substitution (4.3.6) is calculated manufactured goods price equilibrium. But if manufacturing has the number of skilled workers, the supply line will move above right, as the relative supply curve is in the shape of a ladder; paying only works with the demand curve in the lower part of the price of manufactured goods. In this situation, some of workers in the manufacturing technology will be engaged in food work. Similar, if the economy has relatively fewer skilled workers, the two curves will submit to the higher price of manufactured goods. Manufacturing goods prices all expressions:

$$p_M = \begin{cases} \dfrac{1}{A(1-\theta)} & \dfrac{\mu}{1-\mu} \dfrac{2 - L^m - L^{m*}}{A(L^m + L^{m*})} > \dfrac{1}{A(1-\theta)} \\ \dfrac{\mu}{1-\mu} \dfrac{(2 - L^m - L^{m*})}{A(L^m + L^{m*})} \dfrac{1-\theta}{A} \quad \text{if} & \dfrac{1-\theta}{A} \le \dfrac{\mu}{1-\mu} \dfrac{2 - L^m - L^{m*}}{A(L^{m*} + L^m)} \le \dfrac{1}{A(1-\theta)} \\ \dfrac{1-\theta}{A} & \dfrac{\mu}{1-\mu} \dfrac{2 - L^m - L^{m*}}{A(L^{m*} + L^m)} < \dfrac{1-\theta}{A} \end{cases} \tag{4.3.7}$$

where p_M is the relative supply curve (Fig. 4.4).

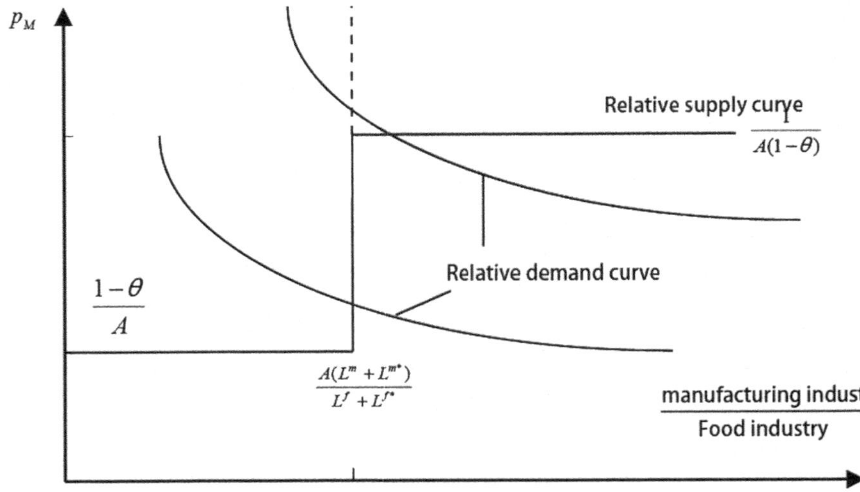

Fig. 4.4 Demand and supply relative to the food industry in the manufacturing sector

The relative price of manufactured goods is determined by the relative price of workers, which determines their pay. If the price is low, workers with manufacturing technology will work in the food department.

In the static model, specifically a population professional transformation cannot be determined because the external dynamic model and local conditions will decide in each region of the manufacturing sectors of employment. Desmet (2000) further studied the conditions in which human capital can flow, a special dynamic process. The Desmet (2000) model involves nonlinear analysis, so we do not analyze it in detail (the reader can refer to the original). Desmet (2000) states the following corollary:

If $(\theta(1-\mu\theta))/(1-\mu) > \beta > 0$ and $1/2 < \mu \leq 1/(2-\beta)$, there are three possible steady states: one is the production of two kinds of products, the other two are the two fully professional regions. And, if $L^{m*}(0) > L^m(0)$, then the formation of the western region is manufacturing specialization, and the eastern part is the food industry. But if $L^{m*}(0) < L^m(0)$, then it is the opposite.

Here θ is the technical gap between technology and the food industry, for the consumers in the production of consumer spending preferences, and for the lifetime of the actual work.

From a given parameter, there are three possible steady states: one is internal homeostasis, the other is the two regions that are fully professional, resulting in the differentiation of the two regions, so that the economic development of the economy is not balanced.

Further discussion also shows that the West has formed a complete specialization of manufacturing technology, and the East food specialization: the relative utility or per capita income is

$$\frac{U^*}{U} = \frac{\mu}{1-\mu} \tag{4.3.8}$$

Because $1 > \mu > 1/2$, it is clear that the per capita utility (income) in the West is higher than that of the East. Despite the high cost of manufacturing, a new generation of East China continues to be engaged in food technology because of the higher costs of manufacturing technology.

In this inference, the author introduces a concept for (after depreciation) real income; the $W_M(t)$ of the model is an interesting aspect in that the output is completely determined by the history. The workers' decision rule is that the eastern part of the people who were born in $W_M(t) > 1/(1-(1-\beta L_M(t)))$ will choose manufacturing technology; the Western partner if found as $W_M(t) > 1/\left(1-\left(1-\beta L_M^*(t)\right)\right)$ will also choose to manufacture.

Desmet (2000) demonstrated that technological innovation is the positive force in economic development. In the process of regional specialization, technological innovation is constantly occurring; a direct problem is that this technological innovation will have impact on what has occurred, but especially in backward areas it cannot because of the professional choice of direction, so that the position of the economic backwardness has been reversed; but Desmet also discussed this issue.

To discuss this problem, Desmet (2000) considers that a regional economic system introduces a new technology of exogenous manufacturing, and the benefits of new technologies are greater than the old ones, that is, $A' > A$. We can explain this technological innovation from two aspects. It is a new technology that can produce the same products, such as small factories instead of large iron and steel factories; it can also produce new technologies that can replace the original products, such as plastics and composite materials to replace steel. Desmet (2000) believes that the East or West using this new technology is determined by two factors: the current level of wages and the two regions of the new technology adaptability; in short, development will be subject to the lock. On the one hand, the new industry can be drawn to the low wage areas; on the other hand, the rich region is more in need of new technology, in this case, although the wages are high; but the West will adopt new technology.

Figure 4.5 shows how a particular professional technique is adapted to the new technology. The original manufacturing and food technology is θ, which is the result of the use of food technology in the original manufacturing sector; productivity will decline by θ. The productivity of new manufacturing after adopting new technology is the greatest, and it is different from the original two. The productivity of the original manufacturing system and the food industry will depend on the distance from the original technology to the new technology, θ'. It is assumed that the distance between the new technology and the old manufacturing technology is close to the level of food technology, as shown in Fig. 4.5. Therefore, the productivity of the workers with the old manufacturing and food technology is $A'(1-(\theta-\theta'))$ and $A'(1-\theta')$. Here, the technology distance means the cost of learning.

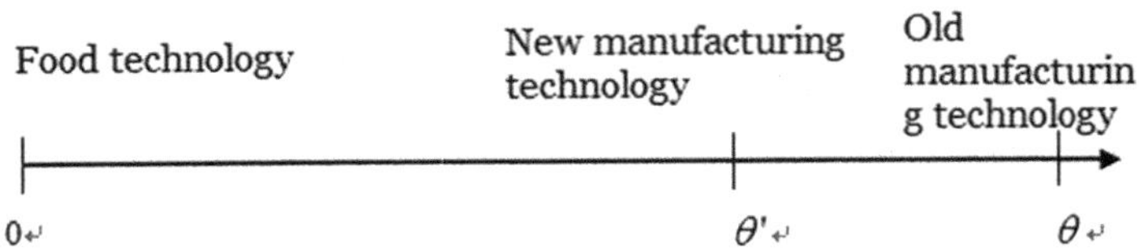

Fig. 4.5 Technology distribution

With the introduction of new technology, the West has formed a specialized production technology, while the East is food technology specialization, namely, the eastern and western sectors of the labor productivity.

	East	West
Food department	1	$1-\theta$
Manufacturing department	$A'(1-\theta)$	$\max[A, A'(1-(\theta-\theta'))]$

This is easy to understand, if

$$A'(1-(\theta-\theta')) > A \tag{4.3.9}$$

That is, in the West after the cost of learning, labor productivity is greater than the original labor productivity when using the new technology.

From a practical point of view, if the old manufacturing technology and new manufacturing technology are closely related, namely, $\theta - \theta' \approx 0$, the West will turn to the new technology. When the (4.3.9) type is not established, the West will continue to use the old technology production and processing products.

On the other hand, if workers in the eastern part of the new technology can get higher wages, then the East may be part of the formation of manufacturing specialization.

$$p_M A'(1-\theta') > 1 \tag{4.3.10}$$

In theory, the price of a particular product depends on whether it is a new technology, that is, the price is

$$p_M = (\mu/(1-\mu))(1/(\max[A, A'(1-(\theta-\theta'))])) \tag{4.3.11}$$

This equation allows us to rewrite the eastern part of the new technology development manufacturing conditions (4.3.10):

$$A'(1-\theta) > \frac{1-\mu}{\mu}\max[A, A'(1-(\theta-\theta'))] \tag{4.3.12}$$

This solution means that if wages in the East are lower than in the West ($(1-\mu)/\mu$), or if food and new manufacturing technologies are closely related (θ' smaller), the East will adopt new technologies, develop manufacturing, and turn to manufacturing specialization; otherwise it will remain fully professional. In practice, we often see some backward areas resulting from the active use of new technologies, while the use of their low wages, in the developed regions of the manufacturing industry, and the development of competition, has continued to maintain low wages, a new specialization, and rapid economic rise. Some so-called new industrialized countries (NICs) are such. It is also easy to find that if $A'(1-\beta) > A$, then in the Eastern or Western conditions no one is willing to choose the old manufacturing technology.

If the new technology is not related to the old technology, $\theta - \theta' \approx 1$, the condition (4.3.9) is not established. Back to steel, the workers in the West are not seeing any of the benefits of the new technology. However, with the assumption that wages in the backward areas are low enough, new technology (Eq. 4.3.12) will be used in the East. If the external nature of the initial location allows the formation of new manufacturing technology in the East, the market will be snatched from the West, and the West will have to turn to the food industry. In a certain period of the process, the East will be greater than the West, and the characteristics of the regional specialization provide hope for the development of the backward countries. So far backward countries or regions often have low wages and there is no early industrial transformation cost advantage in the use of new technology to catch up with developed countries or regions. This phenomenon is known as "after the advantage."

4.3.4 Regional Integration

After regional specialization, the region's trade and the connection between the regions have been strengthened, and further strengthening of these kinds of connections is required for further cooperation between the regions, which has caused regional alliance and integration. Alliance is an early stage of integration, the formation of the interest community, especially in several countries or regions. The regional integration is the process of forming a unified economic system, political system, and ethical system. However, here economic geography only discusses economic integration.

On regional integration, China from the history of the Warring States period to the Qin Dynasty is a typical example. Actually in the spring and autumn period are the vassal states involved in China's ongoing war; however, when the war is predatory, Mencius said, "spring and autumn no wars." But in the Warring States period, the war was basically annexation, and the expansion of trade was the latter part of the general phenomenon of the late Warring States period. In the last part, the United States with military means could unify the six countries, but its background is difficult to separate from the trade requirements at that time. Lü Buwei was the first to start a unified national trade. The seven countries were unified by military means, but military unity is not the new political requirement. The demands of the economy and trade requirements make the first emperor of the

Qin Dynasty put forward the "car with the track," and "hutong wen," setting the system of prefectures and counties at the end of the Qin because of the terrorism rule. After the peasant uprising, the country once again split; Liu Bang reunified China, but conforming at that time to the warlords, he launched the system of prefectures and counties and the system of enrollment for mutual survival. At the beginning of the Chinese government, to actively implement the "Huang-Lao philosophy," inaction encouraged commercial ways and promoted the development of trade. Then, the division of economic basis was finally destroyed, in the period of the Emperor Liu Che, and with Confucianism formed a unified system of ethics and the cultural system. In many of the tombs of the Han Dynasty unearthed in Xinjiang, we can see a large number of tomb murals written in Chinese characters, and people wearing the same dress as the Central Plains people; a unified multi-ethnic state has been formed in trade development after the development of the ethical system and cultural system. Actually, at this time, from the development of the Western regions the Silk Road opened for trading; without the occurrence of early Western and Central Plains in the unity of the trade, how would the Silk Road have ever occurred? In modern times, there is much regional economic alliance, and this economic alliance with political alliances, at the beginning of this century, put forward the concept of European economic integration. Ultimately, the ideal is to be transformed into the European Union and the euro in the end of the twentieth century. However, in many countries, the economy is undeniably becoming more and more integrated. The most powerful trend in the world is global economic integration; this economic integration is the inevitable trend for regional economic development.

We believe that the integration of the regional economy consists mainly of two parts: one is the formation of the common market characterized by free trade, and the other is the formation of the urban system with an industrial center. A unified spatial structure and industrial system will form under these conditions, and the resources and environment structure can also evolve, but it needs the active factor or common market and urban system; moreover, the two have some independence. At present, in understanding regional integration the main economic theory is about the common market, as discussed by Molle (1997). In contrast, the formation of the urban system is still lacking in theory. Discussions of new development of the urban system (such as Eaton and Eckstein 1997), or the formation of industrial centers, as in the fourth chapter of the Krugman model and the recent paper by Lyons et al. (2001), do not provide a very good description of the role of urban and regional integration. Therefore, we focus on the common market theory.

For formation of the urban system, Braudel (1994), in a study in France, found that the town exists only because of the size of the surrounding villages, and its size is not a fair exchange of wealth (p. 127); the town is the product of trade. Braudel then found that the formation of the city is different: the first is to defend the function of the wall (Wai) in space, although the occurrence of the aggregation is "but the first rule of the city is the." Later, because of the surrounding towns, it becomes a big market. In particular, some cities appear "regularly at every 7 days" (p. 143). We know that the 7-day market means $K = 7$ in the Chris taylor structure, which is a kind of structure to help administrative governance. It is here that the conclusion is consistent with our discussion in Sect. 3, where we point out that the city and the towns are different from the original concept, and the city has evolved. We analyzed the evolution of the urban system in the Yangtze River Delta, the Tang Tianbao first year (742), Northern Song Chongning at 35 years (1102), Qing Jiaqing at 35 years (1820), and in 1992 in the Yangtze River Delta city: population data were calculated by the Gini coefficient in the four periods, the 0.14342, 0.23156, 0.20888, and 0.39683. In other words, with the economic development of the urban system to promote the internal level, the central city is becoming more and more prominent, forming a complete functional distribution.

That is to say, in regional integration, regional economic integration and political integration complement each other; it is also for this reason that driving the EU established the European Parliament. China established a system of prefectures and counties. In addition, Sect. 3 shows regional integration from the early development of their own progress. Early development, specialization, and regional integration are the basic process of regional evolution.

According to the economist's view, economic integration has a broad sense and a narrow sense. In broad economic integration, namely, world economic integration, the economies of the world open to each other mutually, to form a mutual connection, as interdependent organisms. Narrow economic integration, namely, regional economic integration, refers to two or more than two countries or regions, in a government-authorized organization with a super-country of common institutions. Through the formulation of a unified internal and external economic policy, fiscal and financial policies, etc., to eliminate barriers between countries, we can achieve mutually beneficial and coordinated development and resource optimization, and ultimately form a political and economic coordination similar to the process in the United States. This argument has a clear political economy of color. We here say that regional economic integration refers to the existence of regional specialization to ensure the effective operation of trade, investment, and employment and the establishment of a unified economic system, and thus contributes to the formation of unified or mutually beneficial (at least nominally mutually beneficial) economic policies, the formation of a unified economic community. As a precursor to economic integration, regional integration is bound to occur. Several countries and regions are based on geopolitical and economic reasons to establish economic alliances, such as ASEAN and China's so-called Southwest Economic Cooperation Zone; of course, some of the union is not necessarily the performance of economic integration, such as the organization of petroleum-exporting countries, but rather they are a market competition in the interests of the community.

The background of economic integration is the existence of trade, so the ultimate goal of economic integration is the elimination of trade barriers. Therefore, the economist accordingly removes the trade barrier delineation of the extent of economic integration, mainly in the following forms: preferential trade arrangement, free trade area, customs union, common market, economic union, complete economic integration, and regional economic integration after World War II, and, especially after the end of the Cold War, rapid development. Many economists have studied and explored it and put forward many theories and theories. We mainly introduce the theory of a Customs Union and the common market theory. These theories are the new classical economics, which indicates that the theory of regional economic integration needs to be further developed. Although we here discuss regional economic integration, in accordance with the practice of classical economics we always use the "state" to replace the "regional." Integration of the global economy is in development, despite the recent emergence of countercurrents. At present, the emergence of a countercurrent still lacks theoretical explanation. It seems that for those who have become the most powerful through international trade, the performance of global integration is a conflict; obviously this requires regional science theoretical analysis.

On the theory of regional economic integration, integration of trade freedom will bring about the improvement of regional benefits. The theory is based on Smith-Ricardo's theory of comparative advantage of continuation; the book's content is mainly based on Molle (1997).

In international trade, two countries (A and B) were produced by two kinds of products (x and y), and the two production elements were labor (L) and capital (C). At the beginning, these two countries are two closed economies. They have different production factors, which means they have different production costs. Figure 4.6 is a national A and a national B. From the perspective of the producer, the curve that is concave to the origin is the so-called "production possibility" or "deformation" curve, which means that a certain number of production factors occur in the production of X products and Y products of a different number of combinations. Because these two countries produce essential factors and technology, their curves are different from the consumer's point of view: convex (to the origin) curve is the indifference curve, and the combination of X products and Y products is the same. Because of the different conditions, the two countries have no different curves. When the no-difference curve and the deformation curves intersect, production and consumption will occur. The price of the products rate (and the tangent of the abscissa intersection of alpha and beta) are also different.

We assume that these two countries have trade relationships; each country only needs relatively few of the production factors of the investment to be able to carry out the professional production of a certain product. Until the price of the two countries is the same, specialized production will stop. In the nation A, because of the production of more x and less y, production goes from the I_A point to the H_A point. In country B, because of the production of more y and less x, production is transferred from I_B to H_B. From the indifference curve, the two countries can achieve a higher indifference curve, such as a trade while increasing the welfare of the two countries. Because of the change of trade and price, there is no difference curve and all the original production possibility curves are without intersection.

From the supply and demand curves of Fig. 4.6 (A), we can see the production and trade of the two countries; the number of x products in the national OD, the numbers of its own production, and the number of DC is imported. In country B, the number of O_BD1 is for their own consumption, and C1D1 is for export. Therefore,

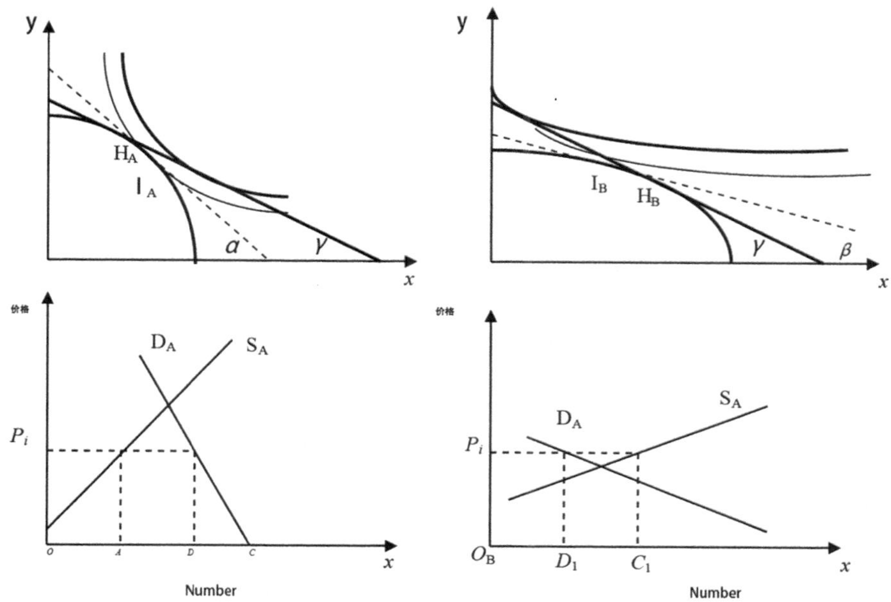

Fig. 4.6 The advantages of regional economic integration

the common improvement of CD = C1D1. These benefits are a way to achieve higher production through lower production and higher production costs.

After the integration is the union of the first regional customs. The advantages of the customs union are mainly the following aspects:

1. Economic benefits: the establishment of the customs union to strengthen the market competition among member countries, the degree of specialization, the use of more efficient production resources after the establishment of a more efficient customs union, the promotion of commodity circulation, can strengthen competition, break the isolation; economic benefits can, therefore, be improved.
2. To obtain economies of scale: after the establishment of the customs union, member states of the domestic market to the United States, the free market expansion occurs, so that member countries can obtain professional and economic interests.
3. Stimulating investment: the establishment of the customs union, the expansion of the market, greatly improving the investment environment, attracts member countries to expand investment, and also can attract nonmember countries to the member countries; to improve product competitiveness, and improve product quality, reducing production costs will increase investment. To nonmember countries, to obtain the benefits of the elimination of customs duties, to break through the Union member state of the discriminatory trade measures, will expand the investment to improve the competitiveness of their own manufacturers.

4. To promote technological progress: after the establishment of the customs union, because the elements of production in the member countries can move freely, the market tends to be unified and the competition intensifies; the investment scale expands, promoting research and development, and technology progress, and accelerates the economic development of the member countries.
5. Improving the liquidity of the elements: the market tends to be unified, the production factors can move freely among the member countries, so it is possible to improve the liquidity of the elements, promote the rational allocation of the elements, and reduce the possibility of idleness.

The further development of the customs union is the establishment of a common market. The common market and the customs union are different, and the integration level is a step further than the customs union. The purpose of the common market is to unify the small market, which is divided by the protection of nature and, with the combination of the big market, through the fierce competition in the market, realize the interests of mass production and so on. The core of the common market theory is that it is possible to obtain scale economy by expanding the market, to realize the technical interest.

If common market conditions can develop to achieve the tariff of the Union, the state, and the country, if tax barriers exist, it is impossible to have a common market. There is no tariff in theory on the area of a country. In addition, it is necessary to produce the elements of the labor and capital in which the capital can move freely. At present, the development mechanism has been analyzed (Molle 1997) from the labor market, the capital market, and the agricultural product market, which is the common market: labor market integration is a factor of market integration. In general, the market integration of other factors, which is not only beneficial to economic development but also the production of other factors, will make the redistribution of capital and labor better; however, this kind of allocation may lead to regional disparities.

It is clear that the book is based on Molle (1997) to discuss or explain the integration of a region; however, regional science needs to develop a theory of regional economic integration. In history, the Chinese economy has experienced a number of integrations and splits. The history of the split may be caused by war, but can also happen in local areas to maintain their economic advantages and work against the integration. For example, developed Shanghai is not keen to integrate with the East China region; and the leading economy in Guangdong, because of the need for resources and labor, has actively called for the integration of China's southern economy. This in turn requires regional scientists to study the historical phenomenon in depth.

4.3.5 Convergence of Region Growth

In the development of a region, industrialization, specialization, and the final trend of economic integration are expected; the problem lies in the backward areas, and countries in the process of integration will not catch up with the developed countries in the process of the per capita income level. In fact, in industrialized countries, growth in the last two centuries is much larger than before, but in many countries in the Third World, there is no obvious change.

According to the theory of new classical economics, the marginal effect decreases, the economic growth rate will fall in the developed regions, and the regional growth rate should be generally greater than that of the developed regions, with the convergence of the economic level. Observation of the history of the United States in the 100 years of development shows that in the United States the economy has convergence: "Poor countries tend to grow faster than rich countries in per capita terms, and converge to the same steady state, without any other features of the economy." That is to say, the low level of the initial economic area has a higher per capita growth rate, so after a period of time, the backward areas will catch up with the developed areas, and eventually they will have the same and stable development, that is, achieve the same steady state.

However, this status is not true in the world. During the period of 1960–1990, the average annual per capita income of the state during the year was always between a negative value and 7%. The exception is the newly industrialized countries in Southeast Asia, including Japan, which has experienced a remarkable growth period. After World War II, the developed market economy makes a per capita real annual income of about 3%. Western European countries, the United States, Canada, Japan, and Australia make up this group. The early period of the most developed countries, the United States and Switzerland, is below the average growth; some underdeveloped countries are well above the rate of growth. The former "Oriental Group" has a much lower per capita income, and the growth rate is much lower. Records show that after the war, the economy has just recovered, and during 1975–1990, the economy was still very depressed in many South American countries. On the other hand, Asia is in the beginning of this period of low income, only gradually achieving a high growth rate. Recently, China's economic growth is particularly significant, but the level of development is still low. In India at the same time are seen a lower level of income and lower growth rate; in Africa, especially in the second half of the same period, 1975–1990, the real rate of many countries has been reduced. Today, the per capita income of sub-Saharan Africa is the lowest.

In economics, "stylized facts" describe many observations of the economic situation; for world economic growth, Kaldor presented below the top six stylized facts to which other authors, such as Bresthchger (1999), have added.

- The per capita output has continued to grow, and the growth rate has not declined.
- Per capita, material capital continued to grow.
- The rate of return on capital is almost stable.

- The material capital–output ratio is almost stable.
- Labor and capital are almost stable in national income.
- There are differences between the per capita outputs in different countries.
- In the horizontal comparison, the population growth rate is negatively related to the growth rate of per capita income.
- The high proportion of the total labor force in the objective influence of the technical labor.
- At the beginning of a period, there is no simple connection between the level of income and the rate of subsequent growth.

These facts show that the convergence of regional growth does not necessarily exist, and the world development forms different economic levels of the national group in a country; it is been divided into different levels of economic regions, such as Eastern China and Western China. All these ideas indicate that economic growth may converge to different levels, that is, the regional economic growth path has multiple equilibrium states. Barrow and Sala-i-Martin have studied the concept of conditional convergence, the main idea being that an economy farther apart from its steady state experiences faster growth; that is, backward areas are not necessarily better than developed regions in faster growth as their growth rate is conditioned on their respective steady-state positions. This equilibrium is different, and as we all know regional growth clumps in the presence of different convergence. Convergence analysis is still mainly dependent on statistical findings. According to the theory of Barro and Sala-i-Martin (2003), using the model to the fitting of a regional set, we can find the convergence of the region. The model is as follows:

$$Dy_i = a_0 + a_1 y_{i,t0} + \varepsilon_{i,t0} i = \overline{1,m} \tag{4.3.13}$$

Type a_0 is the intercept term, $\varepsilon_{i\,,t0}$ is the error term, Dy_i is the i regions (countries, regions or provinces, municipalities, and autonomous regions) per capita GDP (gross domestic product) average annual growth rate, $y_{i\,,t0}$ for i region at the initial time, and t_0 of the per capita numerical GDP. Subscript i is for the i region of encoding, M is the region number, the subscript t_0 is the initial time t_0. (4.3.13) model, a summary of the OECD national situation. In the Wang et al. (2004) analysis of two factors, the level of investment in infrastructure and population growth rate, the model is as follows:

$$Dy_i = a_0 + a_1 y_{i,t0} + a_2 (I/Y)_{i,t0} + a_3 RP_{i,t0} + \varepsilon_{i,t0} \tag{4.3.14}$$

$y_{i\,,t0}$ is the initial time $t0$ of the region, the per capita GDP of numerical, $(I/Y)_{i\,,t0}$ is the initial time $t0$ days, the total investment of infrastructure of its GDP ratio of the annual average value, $RP_{i\,,t0}$ for the initial time of $t0$ at i provinces, municipalities, and autonomous regions, and the average annual population growth rate, where a_0 is the intercept term and $\varepsilon_{i\,,t0}$ is the error term.

Equation (4.3.14) can find the existence of divergence, but the convergence and convergence points of the major regions are ignored. The new economic theory study found that the steady-state position of each region is feasible. The growth rate used in the regression is actually an average of time, and the average value of the long term may neglect some important changes that occur during this period. For this purpose, the Funke and Strulik (1999) method is used to explain the multiple equilibrium state in the economic system, and even more important is the ability to estimate the relative steady-state position in each region. To this end, make the $Y_{i,t} = y_{i,t}/y_t^+$, where $y_{i,t}$ is the i district, at the moment t per capita GDP, and $y_t^+$is at the moment t national per capita GDP. Considering the different steady state, the $Y_{i,t}$ expression is

$$\ln (Y_{i,t}) = vY_{i,0} + \rho_i \ln (Y_{i,t-1})\, i = 1 \ldots \mathrm{m}; \quad t = 1, 2 \ldots \mathrm{T} \tag{4.3.15}$$

The model considers the impact of initial GDP on future growth by parameters, and the corresponding statistical model is written:

$$\ln (Y_{i,t}) = \alpha_i + \rho_i \ln (Y_{i,t-1}) + \varepsilon_{i,t} \tag{4.3.16}$$

where $\varepsilon_{i,t}$ is an error term of mean zero and variance is σ_ε^2. On the basis of Funke and Strulik (1999), the $Y_{i,t}$ (relative) steady-state value is $\alpha_i/(1 - \rho_i)$.

The foregoing analysis of the convergence of the regional economic development of our study does not give a change in the process and extent of economic development, so we need a new approach to solve this problem. The calculation method of the index of Searl is as follows (Terrasi 1999):

$$IC = \sum_i y_i \log(y_i/x_i) = IC_{br} + IC_{wr} \tag{4.3.17}$$

$$IC_{br} = \sum_r Y_r \log(Y_r/X_r) \tag{4.3.18}$$

$$IC_{wr} = \sum_r Y_r \left[\sum_i (y_i/Y_r) \log(y_i/Y_r/x_i/X_r) \right] \tag{4.3.19}$$

IC is the form of total inequality, IC_{br} is the regional inequality, IC_{wr} is the region's unevenness, and y_i and x_i are the GDP, respectively, for the provinces, municipalities, and autonomous regions, and Y_r and X_r are the population accounted for two of the total values of the region, h and F for each region accounting for the share of the value. With the Searl index, we can study the inequality of regional economic development and the change of its spatial structure.

4.4 Regional Spillover

4.4.1 Concept of Regional Spillovers

The concept of spillover, as the positive content, first appeared in an article by Grossman and Helpman (1991). Grossman and Helpman (1991) pointed out that in a country a region of knowledge spillovers will lead to the economic development of this country, this region. Spillover also can be a curse, that is, the impact may be good or bad. For example, a large number of members of the Chinese Academy of Sciences focused on Zhongguancun, which has caused a high technology product trade to develop as a positive spillover. The external non-economy of environmental pollution is negative spillover. There are many forms of spillovers, such as knowledge spillovers, public goods, and GDP spillovers. These spillovers occur in the region, as a geographical phenomenon.

Regional spillover is a kind of multi-regional phenomenon. In the process of regional economic development, as the driving force, not only as the new classical growth theory, the emphasis is on the capital to the population, the region's interaction. In regional interaction, in recent years, it has been found that regional spillover has an important significance. What is it? This concept in the purely economic sense involves two meanings.

First, consider that "capital" as the factor that determines the growth of all types of capital or capital stock (showing the value of the economy) can be considered as a factor of production. Therefore, in addition to material capital such as machinery, equipment, and plants, another important capital form of human capital, knowledge capital, and environment and public infrastructure provided by the government are capital; they have a sense of economic growth.

Second, taking into account the human capital, knowledge capital, environment, and public infrastructure provided by the government can affect the capital outside the area and family, enterprises, etc., which produces a "spillover," such as knowledge; in general, we know that this is not like material capital that produced the "use of endless spillover." Tourism resources, being developed for the National Park, also led to the development of related industries. Facts show that knowledge is spread by culture in the region so that more and more people benefit from the knowledge spillover. However, in fact, there is no such mechanism. Grilichers (1979) considered that there are two kinds of knowledge spillovers: the first spillover is reflected in the output of the product; the second is pure knowledge spillovers, which is not reflected in the product, but through meetings, seminars, and other forms. These two ways are the distinction between pure knowledge spillovers as a kind of information exchange, for which we do not need to pay directly, and the product output for which it is required to pay directly. Generally, the spillover has the same characteristics. In recent years, the concept of GDP spillover is proposed, which is defined as GDP (Peeters 1997): a region that is the result of its fiscal policy, monetary policy, or other exogenous shocks and other regional macroeconomic variables. For example, the increase of German growth will stimulate the growth of Shanghai; Holland's opening will drive economic growth in the Yangtze River Delta—this type of argument. Therefore, the spillover of GDP refers to the change of the GDP of foreign (regional), which is caused by the impact on some of the domestic areas.

Spillover may produce two aspects, positive and negative; positive spillover affects regional economic activities, which are the external effects that promote regional economic progress. Negative spillovers, such as economic growth to bring environmental pollution, the region needs to pay environmental remediation costs, an increase of regional development costs.

The analysis of spillover effect can be traced back to the 1960s. In Mike Doga (MacDougall), in the analysis of foreign direct investment (Direct Investment Foreign, hereinafter referred to as FDI) and the general welfare effect, for the first time the technology spillover effect is considered as an important phenomenon. Later Kodun (Cooden), respectively, investigated the effect of FDI on the best tariff, industrial model, and welfare, which also referred to the spillover effect, but the purpose of the study was to estimate the cost and benefits of FDI, so the real meaning of the spillover theory is too distant. Many of the later studies concluded that the spillover effect came from two aspects: one is that it is a model of imitation and diffusion; the other is competition. The former is the increasing function of the difference in technical information; the latter is mainly determined by the market characteristics of foreign enterprises and local manufacturers and their mutual influence. Caniëls (2000) pointed out that "knowledge spillover is through the exchange of information and access to intellectual achievements, and not to the creators of knowledge to compensation or compensation given to less than the value of intellectual achievements." Knowledge spillover and knowledge diffusion are different concepts, and knowledge diffusion is more extensive. It covers all kinds of the ways that knowledge moves across time and space, and knowledge spillover is just one of them. However, precisely, the spillover and diffusion are essentially different. We observed a phenomenon: a public toilet in Zhongguancun, Beijing Province, with the development of a number of small restaurants nearby. Investigating the reason, it is because the taxi drivers found the toilets here for their help after a meal, so they are willing to choose to eat in the vicinity, to reduce their flying time, knowing the taxi driver must do a lot of work to have a higher income. Although the results included toilet odor diffusion, also the mechanism operated for restaurant benefit. The content of the diffusion itself is not only related to the benefit but also may affect the benefit.

Regional spillovers, overall, have polymorphism. There are four important forms: one is the negative spillover caused by environmental pollution; two are knowledge and technology innovations, the two innovations that led to knowledge spillovers, generally a positive spillover. An important form of knowledge spillover is R&D spillover. The three forms are the GDP spillover of regional economic interaction, and the mutual use and development of regional markets; mutual trade promotion and technological interaction are all GDP spillovers. For GDP spillover, it is possible. For example, China's growth may open up the US market, thus driving the growth of the United States. Environmental spillover and knowledge spillover may be some of the factors that have a positive and negative nature. However, the new economic growth theory point of view places more emphasis on the positive aspects of the spillover.

The fourth kind of spillover is the resource spillover, which is the difference between the regional resources and the advantage of the regional resources, and the public facilities as a kind of resource is a kind of spillover. Grossman and Helpman (1991) proposed that knowledge spillovers have important significance to regional economic growth. Caniëls and Verspagen (2001) emphasized the spillover can be considered as the product of the knowledge economy environment, the regional knowledge spillovers and the most important spillover. Empirical analysis by Wang et al. (2004) found that the knowledge spillover in the region was attenuated by exponential decay. Conley and Ligon (2002) developed a measurement method to measure the regional spillover intensity.

Limited in space, this book mainly introduces knowledge spillovers; more discussion is found in Teng and Wang (2010), in the book *Regional Spillover*.

4.4.2 Regional Knowledge Spillovers

In the case of knowledge spillovers, regional development is complex, which highlights the surface. Ben-David (1998) considers an economic system with J countries and the national I production of goods; each country's products are not the same. Then, we use the following model to discuss the knowledge spillover of the multi-region system.

$$\dot{H}_i(t) = \phi\left[\sum_{j\neq i} a_{ij}\nu_{ij}(t)H_j(t)\right] + (\phi - \delta_H)H_i(t) \tag{4.4.1}$$

Here $H_i(t)$ is the total knowledge stock of the state i at the moment t, and ϕ represents the public productivity parameter, the depreciation rate of δ_H knowledge stock, ν_{ij} is the ratio of i and j, and the ratio to the total output of i, a_{ij} is the knowledge spillover coefficient, which Ben-David (1998) assumed to be constant. In the further Ben-David (1998) hypothesis, by the (4.4.1) type it can be seen if the country does not trade, a country's internal knowledge will grow with the exogenous growth rate B, and then a, the country's knowledge of foreign trade, will not grow. By the (4.4.1) type it can be seen if the country does not trade, a country's internal knowledge will grow with the exogenous growth rate $\phi - \delta_H$, and when $\phi = \delta_H$, the country's knowledge of foreign trade will not grow. As with Lucas (1988), Ben-David assumes that the knowledge accumulation of i in the country has a constant return on its knowledge level, thus assuming that the per capita production function $y_i(t)$ has the following form:

$$y_i(t) = AH_i(t)^{\varepsilon_i}\ \varepsilon_i > 0 \tag{4.4.2}$$

So, the problem of economic growth has been changed to the problem of knowledge growth. To estimate ν_{ij}, and guarantee the balance of trade and market clearing, Ben David (1998) assumes that $L_i(t)$ is the population number in country i at time t, n_i is the population growth rate, and is exogenous, and $c_{ij}(t)$ is the national i at moment t in the commodity j per capita consumption. The commodity i is the valuation standard, then $p_1(t) = 1$, $p_i(t)$ is the price of i commodity t, and τ_{ij} is national i on the national j import tariffs (by definition, $\tau_{ii} = 0$). Here the tariff is given, and the constant. And every consumer in the country I has the same preference: the national i budget constraint is

$$\sum_{j=1}^{J} \frac{p_j(t)\cdot\left(1+\tau_{ij}\right)}{p_i(t)} c_{ij}(t) = AH_i(t)^{\varepsilon_i} + g_i(t) \tag{4.4.3}$$

Here

$$g_i(t) = \sum_{j\neq i} \frac{p_j(t)\tau_{ij}c_{ij}(t)}{p_i(t)} \tag{4.4.4}$$

ν_{ij} is defined as a measure of the degree of openness of the state i to the state j:

$$\nu_{ij}(t) = \frac{\frac{p_j(t)}{p_i(t)}L_i(t)c_{ij}(t) + L_j(t)c_{ij}(t)}{L_i(t)y_i(t)}, \quad i\neq j \tag{4.4.5}$$

After deduction

$$c_{ii} = \alpha_{ii}Q_i y_i \tag{4.4.6}$$

$$c_{ij} = \alpha_{ij}\frac{p_i}{p_j\left(1+\tau_{ij}\right)}Q_i y_i \tag{4.4.7}$$

$$Q_i = \frac{\prod_{j\neq i}\left(1+\tau_{ij}\right)}{1+\sum_{j\neq i}\tau_{ij}\left(1-\alpha_{ij}\right)+\sum_{j\neq i,k}\sum_{k\neq i,j}\tau_{ij}\tau_{ik}\left(1-\alpha_{ij}-\alpha_{ik}\right)+\cdots+\left(1-\sum_{j\neq i}\alpha_{ij}\right)\prod_{j\neq i}\tau_{ij}} \tag{4.4.8}$$

The price obtained from market clearing conditions

$$p_i = \pi_i\frac{L_1y_1}{L_iy_i}\ i = 2, \ldots, J, \tag{4.4.9}$$

Here π_i for all I and J $(i\neq j)$ is a function of $\widehat{\alpha_{ij}} = \frac{\alpha_{ij}Q_i}{1+\tau_{ij}}$. If $J = 3$, then

$$\pi_1 = 1 \tag{4.4.10a}$$

$$\pi_2 = \frac{\widehat{\alpha}_{12}\left(\widehat{\alpha}_{31} + \widehat{\alpha}_{32}\right) + \widehat{\alpha}_{13}\widehat{\alpha}_{32}}{\widehat{\alpha}_{21}\left(\widehat{\alpha}_{31} + \widehat{\alpha}_{32}\right) + \widehat{\alpha}_{23}\widehat{\alpha}_{31}} \tag{4.4.10b}$$

And

$$\pi_3 = \frac{\widehat{\alpha}_{13}\left(\widehat{\alpha}_{21} + \widehat{\alpha}_{23}\right) + \widehat{\alpha}_{12}\widehat{\alpha}_{23}}{\widehat{\alpha}_{21}\left(\widehat{\alpha}_{31} + \widehat{\alpha}_{32}\right) + \widehat{\alpha}_{23}\widehat{\alpha}_{31}} \tag{4.4.10c}$$

One of the most important problems in the Ben–David model is that knowledge spillover is considered a constant, and the fact is that knowledge spillover is not a constant but a function of the knowledge gap. Caniëls (2000) and Caniëls and Verspagen (2001) noted that the knowledge gap effect of knowledge spillovers is a nonlinear function form. Pyka (1997) proved that the node states in the formal network satisfy the exponential function according to the Markov property of the time transfer:

$$p = \alpha\, e^{\beta X} \tag{4.4.11}$$

Here, X is the state difference, which is knowledge spillover intensity. Wang et al. (2004), according to the actual data, proved that the spillover effect of the knowledge gap in the form of the index is satisfied with the characteristics of spatial interaction. It is reasonable, so we use the following forms of spillover:

$$a_{ij}(t) = \begin{cases} e^{-\left(\frac{1}{\delta_i}\ln\frac{H_j(t)}{H_i(t)}\right)^2} & H_i(t) < H_j(t) \\ 1 & H_i(t) \geq H_j(t) \end{cases} \tag{4.4.12}$$

We use the piecewise function form because from the actual situation, we believe that when the stock of knowledge in i is greater than that of i, j can absorb the knowledge of digestion from the j countries (such as the ability of the United States to absorb the knowledge of China). In the case of simple policy convergence, there is no difference between the regions, so we can no longer use the δ_i to make the difference, and the $1/\delta$ defines the knowledge spillover and diffusion of the regional network.

This model describes the nonlinear relationship between trade and the knowledge network, which can reveal the complexity of the system, but it is difficult to give a solution of the system, so it can only be studied by numerical simulation.

First, we need to further explain the parameters of the Ben-David model. Here, we assume that the global system is of three worlds: the least developed countries, moderately developed countries, and developed countries (hereinafter referred to as the poor and middle-income countries; Wells Fargo). Wang et al. (2007) carry out numerical simulation for free trade in two different ways (unilateral tariff reduction, bilateral tariff cut). Two simulation parameters are set as follows:

Parameters of tariff reduction

To simplify, similar to Ben-David (1998), some parameters as set to three countries are the same, and specific parameters are given as follows: $H(0) = (1, 5, 10)^T$, $L(0) = (1, 5, 1)^T$, $A = 1$, $\phi = 0.1$, $\delta_H = 0.05$, $\rho = 0.04$, $n_i = 0.02$, $\varepsilon_i = 0.3$, $\alpha_{ii} = 0.6$, $\alpha_{ij} = 0.2$ $(i \neq j)$, and $\tau_{ij} = 0.75$. By setting the parameters, it could be seen that the state is the initial state of the poorest of the countries, that country 3 is for the first time the richest country, and country 2 is the initial medium-rich country; it has a large population of features. In addition, it is assumed that the utility of consumers in various countries is the largest when consuming the goods of their own country. Tariff is set to 0.75: such a high level is to make the image more easily studied. In unilateral tariff reduction, we assume that of the medium-rich countries from the beginning of the fourteenth period of tax cuts, similar to China's entry into the World Trade Organization (WTO) policy, the national 3 (rich) of the import tariff is cut to 5 in the 0 period. At the same time, all the other duties remain unchanged.

Parameters of bilateral tariff reduction

After entering the WTO, more countries are facing bilateral tariff reduction. In this context, we take most of the parameters such as unilateral drop, but in this study, we have two parameters that were modified.

First, in tau fell 23 based, we assume that in 2 of the 3 countries the tariff also from the 14 start points (5 average) is reduced to 0. That is to say, in 19 (of 19), the state and national 3 completed removal of the bilateral tariffs.

Second, on the basis of the previous, we assume that the country's 2 and 3 of the consumer goods will have a higher utility, that is,

$\alpha_{23} = \alpha_{32} = 0.267$, the corresponding is $\alpha_{21} = \alpha_{31} = 0.133$.

Differing from the conclusion of the Ben-David anti-proportional diffusion model, Ben-David considers that the tariff reduction (i.e., the realization of free trade) will inevitably lead to the steady-state transfer of the developed countries to the less-developed countries, and that convergence of the economies will result. Studies by Wang and others show that, in fact, this is not so. The emergence of the convergence phenomenon depends on the size of the knowledge of the economy, that is, the size of the existing knowledge gap between them. In the foregoing model, the most direct impact is the value of the parameters δ. Therefore, we simulate the different δ values of two kinds of tariff policies. It has been found that the property rights of the developed countries can effectively influence the development of the country. Under the strict control of intellectual property, the gap between the two regions may become more and more larger, and the world economy will not converge. If developed countries relax the regulation of intellectual property, to transfer technology more actively to the developing countries, this will help the world economic convergence (Fig. 4.7).

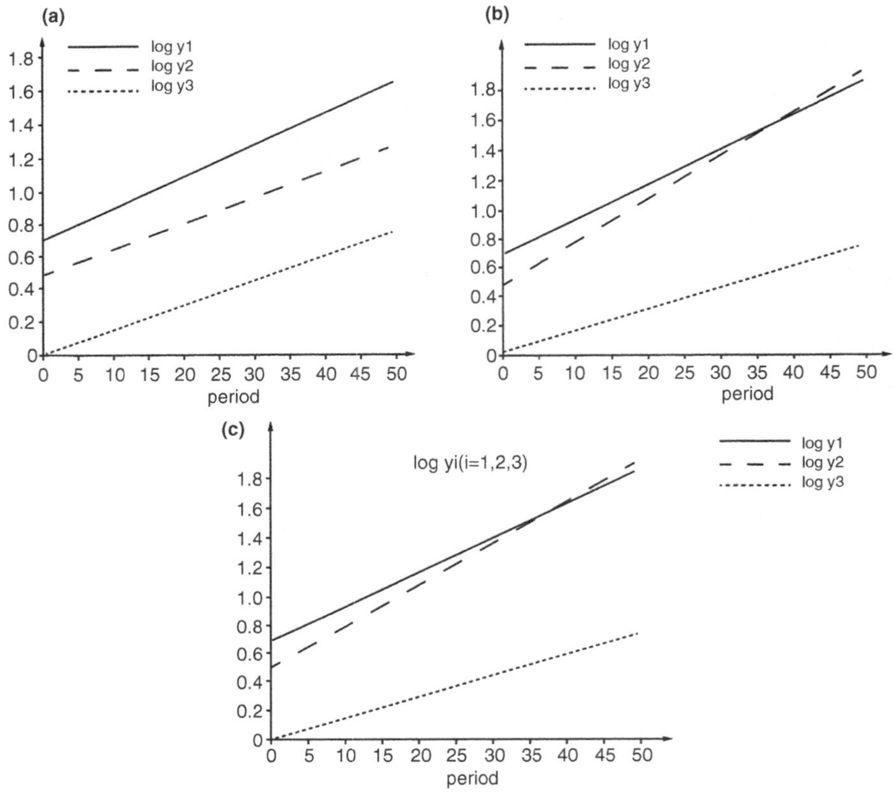

Fig. 4.7 Changes of per capita income in countries with different knowledge spillovers. (**a**) Knowledge spillovers with large damping. (**b**) Knowledge spillovers with medium damping. (**c**) Knowledge spillovers with small damping

4.4.3 Region GDP Spillover

Regional spillover is not only the knowledge spillovers of the region, but also the GDP (gross domestic product) spillover. What is the GDP spillover? In the integrated economy, the national economy of each country exists in a global system that depends on the others. In this system, each country provides consumer goods, intermediate goods, and capital goods with its trading partners, but also relies on trade partners to provide the market for their products. Moreover, it is more and more obvious that the countries still rely on technology transfer to each other, and learn from each other by learning methods, organization models, marketing strategy, and product design. These activities affect the welfare of the country and the growth rate (Bayoumi et al. 1999), which is the GDP spillover.

Douven and Peeters (1998) define the national GDP spillover as a (country domestic) fiscal policy, monetary policy, or other exogenous shocks, resulting in an external state (country foreign) of the economic variables. Or, the GDP spillover results from the impact of a number of foreign GDP changes (Douven and Peeters 1998).

From the nature of economics to describe the spillover, it could be said that in the capital, finance, and the product market with more and more integration process, the interaction between the economies has become increasingly important. The economic behavior of a country will have an important influence on the economy of the neighboring countries. This external effect, that is, the so-called spillover, may be great, especially when the states are closely linked (Peeters and Douven 1998). Therefore, in the theory and the empirical economy, more and more interest is concerned with the measurement, the symptoms, and the size of the spill.

The earliest analysis of international spillover theoretical work can be traced to Mundell (1963) and Fleming (1962), and later others continued to develop their model. For example, for McKibbin and Sachs (1991) and Krugman (1995) the Mundell–Fleming models are static and the two countries in the model are completely symmetrical in parameters and structure. Subsequently, some scholars have carried out an empirical study on the multi-model. Helliwell and Padmore (1985) and Frankel (1988), have carried out a number of comparative studies on spillovers. They used the same impact analysis method in different multi-country models, and compared and analyzed the response of different models. For example, they compared the response of foreign GDP to the financial or monetary shocks that had occurred in their own country.

Analysis of GDP spillovers causes GDP to simulate the tidal current and therefore is responsible for the computerized model; the molding of the prototype is actually the two-country Mundell–Fleming model under Doven and Peeters (1998) on the development of an open economy. The M-L model contains two regions: national and foreign. Each region has a total demand curve, the total supply equation (or Phillips curve), and a money demand equation. The consumer price level has been defined as the weighted average of domestic and foreign prices, and the two regions are linked by interest rate parity.

Table 4.3 introduces the M-F model (Douven and Peeters 1998) in the McKibbin and Sachs (1991) version based on the development of the modified version of the dynamic; it is computable, here the economies of both countries are their economies (no asterisk) and foreign economies (asterisk). The model has a key assumption, namely, that the capital flows completely. The model has four versions, each of which is shown in Table 4.4.

No asterisk appears in the model system here in which the variables are local variables; the asterisked variables are foreign variables. In all the variables, except for the sum of the natural logarithm form, the parameters are assumed positive. The negative value is the lag angle, that in the future (expected value) endogenous variable definition: q = real GDP; i = short-term interest rates; e = exchange rate, defined as a unit of foreign currency prices; λ= real exchange rate; p= price level; and p^c= consumer price level. The exogenous variable definition: m = nominal monetary equilibrium; g = real government spending; T = GDP growth time trend;

Table 4.3 M-F dynamic two-model revision

Model A $\left(\varphi_1 = \varphi_1^* = 1\right)$	
Model B	
$m - p^c = \omega_0 + \omega_1 q - \omega_2 i + \omega_3\left(m_{-1} - p_{-1}^c\right)$	(4.4.13)
$m^* - p^{c*} = \omega_0^* + \omega_1^* q^* - \omega_2^* i^* + \omega_3^*\left(m_{-1}^* - p_{-1}^{c*}\right)$	(4.4.13*)
$q = v_0 + v_1\lambda - v_2(i - p_{+1} + p) + v_3 q^* + v_4 g + v_5 T$	(4.4.14)
$q^* = v_0^* + v_1^*\lambda - v_2^*\left(i^* - p_{+1}^* + p^*\right) + v_3 q + v_4^* g^* + v_5^* T$	(4.4.14*)
$p - p_{-1} = \varphi_0 + \varphi_1\left(p_{-1}^c - p_{-2}^c\right) + \varphi_2\left(q_{-1} - \hat{q}_{-1}\right) + \varphi_3(q_{-1} - q_{-2})$	(4.4.15)
$p^* - p_{-1}^* = \varphi_0^* + \varphi_1^*\left(p_{-1}^{c*} - p_{-2}^{c*}\right) + \varphi_2^*\left(q_{-1}^* - \hat{q}_{-2}^*\right) + \varphi_3^*\left(q_{-1}^* - q_{-2}^*\right)$	(4.4.15*)
$p^c = \rho_1 p + (1 - \rho_1)(e + p^*)$	(4.4.16)
$p^{c*} = \rho_1^* p^* + \left(1 - \rho_1^*\right)(-e + p)$	(4.4.16*)
$\lambda = e + p^* - p$	(4.4.17)
$e_{+1} = e + i - i^*$	(4.4.18)
Model C	
$q = \varphi_0 + \varphi_1(p^c - p) + \hat{q}$	(4.4.15a)
$q^* = \varphi_0^* + \varphi_1^*(p^{c*} - p^*) + \hat{q}^*$	(4.4.15a*)
Model D	
$\Delta(m - p) = \omega_0 + \omega_1\Delta q - \omega_2\Delta i + \omega_3(m_{-1} - p_{-1} - \gamma_0 - \gamma_1 q + \gamma_2 i - \gamma_3 T)$	(4.4.13a)
$\Delta(m^* - p^*) = \omega_0^* + \omega_1^*\Delta q^* - \omega_2^*\Delta i^* + \omega_3^*\left(m_{-1}^* - p_{-1}^* - \gamma_0^* - \gamma_1^* q^* + \gamma_2^* i^* - \gamma_3^* T\right)$	(4.4.13a*)

Table 4.4 Model evaluation

	Rigid price	Long-term output/inflation balance	Potential output	Money demand
Model (A)	Yes	No	Ensure	Cogan
Model (B)	Yes	Yes	Ensure	Cogan
Model (C)	Yes	No	Ensure	Cogan
Model (D)	Yes	No	Random	Error correction

Source: Ghosh and Mssson (1994): *Economic Cooperation in an Uncertain World*

the main means by labor growth, technological progress, and other spontaneous growth, with random interference; and $\hat{q}$ = potential GDP.

For the model calculation, the rationality of the model is necessary, and the model equation (4.4.13) is the standard LM curve. The equation (4.4.14) is described by that (4.4.18) of the actual interest rate, nominal interest rate, foreign GDP, government spending, and tax rates explaining the actual aggregate demand equation. In fact, it is the representative of the open economy in the IS curve: it shows that the total demand of a country (relative to foreign terms) and the real exchange rate is inversely proportional to the relative real interest rate. Equation (4.4.15) explains the price of GDP by using the Phillips curve. Equation (4.4.16) is expressed as a weighted average of the price of goods for domestic production and the price of imported goods. Equation (4.4.17) is the real exchange rate. Equation (4.4.18) is not covered interest rate parity and said capital flows.

It should be emphasized that in the foregoing theoretical model, Eqs. (4.4.13), (4.4.14), and (4.4.18) have strict assumptions. Equation (4.4.13) is the assumption behind the type of the money market equilibrium condition: the money supply is equal to the money demand; Eq. (4.4.14) is the assumption behind the type that is the national income identity: saving equals investment; and Eq. (4.4.18) is that hidden behind the exchange rate fluctuations and interest rate market. These assumptions only are completely under the condition of market economy: various market and clearing functions will meet, even if it is in reality the developed market economy countries. These assumptions also can only be regarded as an approximation and may be completely off. However, we give a theoretical model of the purpose of this theory not to consider that the actual situation of the area is the case, but to derive the relationship between the economic variables that the ideal situation should have.

Here we discuss the four versions of the model, namely (A)–(D). Model (A) is former to exchange rate expected standard M-F model price rigidity and expected inflation is the consequence, so monetary policy in the short term will affect the output. Equations (4.4.15) and (4.4.15*), the expected inflation of the Phillips unit coefficient (a = 1), means that there is no long-term trade-off between output and inflation. Model (B) is similar to model (A), which is not uniform in the correlation coefficient of the expected inflation in the Phillips curve. So the model (B) has a long-term equilibrium (trade-off) between output and inflation. Model (C) is equivalent to the model (B), but with the Lucas supply curve (4.4.15) and (4.4.15*); the price is completely mobile, monetary policy is invalid, so the model is a new classic. At this point, the other three models are very different. All these models assume that the potential output can be simulated by a deterministic time trend. Model (D) is a trend of potential output. We argue that for this, according to the theory of growth (Barro and Sala-i-Martin 2003): it is population growth and the rate of depreciation of the poor. Model (D) is also different from model (B) in two ways. First, the money demand equation is simulated in the form of error correction, but not only by the delay. Second, the GDP price level is used rather than the consumer price level to narrow the money demand. Table 4.4 summarizes some of the characteristics of the four models.

When there is no difference in the exchange rate of a country's regional issues, the price level has an effect on interregional trade. Wang et al. (2003a, b) used the Douver–Peeters (1998) model of population growth rate and urbanization rate of two factors, through the two regions of the price index to reflect the difference between the nominal exchange rate to reflect the purchasing power of different regions, so as to obtain a simplified model of regional GDP spillover:

$$q = \nu_0 + \nu_1\lambda - \nu_2\left(i - p_{+1} + p\right) + \nu_3 q^* + \nu_4 g + \nu_5 r_p + \nu_6 r_c \tag{4.4.19}$$

$$q^* = \nu_0^* + \nu_1^*\lambda^* - \nu_2^*\left(i - p_{+1}^* + p^*\right) + \nu_3^* q + \nu_4^* g^* + \nu_5^* {r_p}^* + \nu_6^* {r_c}^* \tag{4.4.20}$$

In the model (4.4.19)–(4.4.20), the amount of "*" in the model is represented by the amount of "*", and the negative subscript indicates the value of the previous period: q = real GDP; i = real interest rate; p = price index; g = government real expenditure; t = trend item; and RP and RC are the population growth rate and urbanization rate, respectively. This simplified model is the basis of the study of regional economic interaction. It is necessary to point out that this equation is the result of theoretical analysis, not a regression relationship, and a causal relationship between the variables. In the following analysis, we use the least squares method to estimate the parameters, but cannot rely on the significance of the estimation.

In this model (1), the Wang et al. (2003a, b) simulation of each region has a very significant positive impact, that is, the effect of different regions is significantly different, as can be inferred from the west to the east of the east, and the west has a greater impact on the central. Spillover effect is significant, and the regional economic relevance of our country is good, especially in the eastern part where the spillover effect is strong, but for everyone concerned about the problem of the west, the central area to the west of the GDP influence is stronger than the east. (2) There is no obvious regional advantage, weak interaction, or strong interaction between the central and western regions, and the T value is close to 10% of the statistical test, which shows that there is competition in the product price in their trade with the east, the eastern part of the impact. But the impact of the east on their performance is not obvious. (3) In addition to the west, government financial expenditure has a positive effect on the GDP growth in the region. It is difficult to judge the role of the west; the lower interest rates will contribute to the development of the west, but it will be difficult for the west to develop. (4) In the middle and western regions, the nominal interest rate also has the tendency of a *T* value close to 10% of the statistical test, which shows that the lower interest rate mainly affects the central and western regions, which reduces the interest rate to improve the GDP in the west, which has an inhibitory effect on the central region. (5) The urbanization rate and population growth rate are complex, but the eastern and western urbanization rates have positive significance. The central part is not the key to improve urbanization.

4.4.4 Regional Resource Spillover

As the economic growth of the total cost of resources, when the distribution of resources is not uniform, regional resources and trade will inevitably occur: this raises the problem of optimal growth under resource constraints. To this end, Teng Li et al. used the Bretschger model (1999b) to consider the use of capital in the negative spillover concept: the meaning of this negative spillover is the accumulation of capital to the environment, the destruction of the environment, and the reduction of output. At this time, there is this relationship between production functions and resources and capital:

$$Y = K^{\eta} K^{\alpha} N^{1-\alpha} \tag{4.4.21}$$

$$N = N^{*} K^{-\psi} \tag{4.4.22}$$

K is capital, and *N* the natural resources. The author believes that technological progress is beyond the material capital of the study effect, so there is technological progress A = K^{η},η, ψ. In said technological progress level, the value of the capital is less than 1, its meaning is with the increase in capital investment, meaning production-scale expansion, and the production of the environment has a certain self-purification function, so the impact strength coefficient is less than 1. Especially for renewable resources, N* (4.4.22) in the equation in fact "represents" the total amount of renewable resources; when the K increases, pollution damages the natural resources of the regeneration rate, and therefore can be used to use the remaining N. Thus,

$$Y = K^{\eta} K^{\alpha} [N^{*} K^{-\psi}]^{1-\alpha} \tag{4.4.23}$$

At this point, the Brestchger (1999b) model considers that natural resources are also a form of capital investment and thus affect output. The negative effect of capital accumulation on natural resources will ultimately lead to a reduction in economic benefits. However, the Brestchger (1999b) model is for a single region and does not involve the problem of multi-regional interaction. In this paper, the relationship between capital and resources in the Brestchger (1999a, b) model is developed, and the model of regional spillover growth under resource constraints has been established. The spillover effect is secondary growth effect from the other region; the growth cause with the transfer of resources, which we call the resource spillover. It may be considered that the "spillover" is a "free lunch," and the existence of resource transfer trade. In fact, as Grossman and Helpman (1991) pointed out, there is a spillover of trade. Here we understand that spillover is a kind of economic activity, which is the economic output effect of another economic body. To this end, there are two hypotheses: area 1 and area 2. The two regions have some natural resources: the amount of natural resources in a certain time are N_1 and N_2, when the region does not exist between resources and trade; the production function of the two regions are as Eq. (4.4.21) and the relationship between the amount of resources and capital, such as Eq. (4.4.22). For the region 1, if there is a certain price sale ratio of resources, to sell resources will give region 1 direct income (output), and the remaining resources go into production. At the same time, the production function for region 1:

$$Y_1 = {K_1}^{\eta_1} {K_1}^{\alpha_1} [N_1(1-b)]^{(1-\alpha_1)} + PN_1 b \tag{4.4.24}$$

Among them, here b says that the region 1 sold to the region outside the resources accounted for 1 of the proportion of available resources N_1, because of the resources of the unit price of P, so the type (4.4.24) on the right side of the second is the sale of resources to obtain revenue. Other variables and parameters have the index 1, which indicates that the region 1 is variable. The relationship of N–K (4.4.22)–(4.4.24), has this information:

$$Y_1 = {K_1}^{\eta_1+\alpha_1} ({N_1}^* {K_1}^{-\psi_1})^{(1-\alpha_1)} (1-b)^{(1-\alpha_1)} + P({N_1}^* {K_1}^{-\psi_1}) b \tag{4.4.25}$$

To compare the relationship between the sale of resources and the income (output) of these two cases, we have written the production function of the region 1 not to sell, as shown in the formula (7.4.5):

$$Y_1' = {K_1}^{\eta_1+\alpha_1} ({N_1}^* {K_1}^{-\psi_1})^{(1-\alpha_1)} \tag{4.4.26}$$

By comparing the Eqs. (7.4.5) and (7.4.6), it can be determined whether the region 1 obtains more output from the resource trade, thereby defining the spillover of resource trade:

$$\begin{aligned}\Delta Y_1 &= Y_1 - Y_1' \\ &= {K_1}^{\eta_1+\alpha_1} ({N_1}^* {K_1}^{-\psi_1})^{(1-\alpha_1)} \left[(1-b)^{(1-\alpha_1)} - 1\right] + P({N_1}^* {K_1}^{-\psi_1}) b\end{aligned} \tag{4.4.27}$$

Clearly, if $\Delta Y_1 > 0$, region 1 will obtain more income in the resource trade, the positive spillover; or vice versa, the loss of part of the revenue, the negative spillover.

For region 2, the purchase of resources and the sale of the region 1 $N_1 b$, then the region must be paid for the purchase of resources and capital, and production functions can be written as

$$Y_2 = (K_2 - N_1 bP)^{\eta_2+\alpha_2}(N_2 + N_1 b)^{1-\alpha_2} \tag{4.4.28}$$

The *N–K* relationship can be

$$Y_2 = \left(K_2 - N_1^* K_1^{-\psi_1} Pb\right)^{\eta_2+\alpha_2}\left(N_2^* K_2^{-\psi_2} + N_1^* K_1^{-\psi_1} b\right)^{1-\alpha_2} \tag{4.4.29}$$

$K_2 - N_1^* K_1^{-\psi_1} Pb$ points out that region 2 must provide some of the investment to buy resources, so that the production of capital reduction, obviously, the item should be greater than zero, that is, the investment in the purchase of resources cannot exceed the total amount of capital. To buy resources and when compared to the situation, the relationship between N–K (4.4.21) and the type subscript 2 is

$$Y_2' = {K_2}^{\eta_2+\alpha_2}({N_2}^* {K_2}^{-\psi_2})^{(1-\alpha_2)} \tag{4.4.30}$$

From this, we can define the spillover of resource trade to the region 2:

$$\Delta Y_2 = Y_2 - Y_2' \tag{4.4.31}$$

Comparison of region 2 in the purchase of resources, and not to buy the output of resources, if the purchase of resources is greater than the output of the non-purchase of resources, indicating the existence of resources is a positive spillover; in contrast, if the purchase of resources is less than the output of the non-purchase of resources, the region is a negative spillover; in fact, on the market, this time the region is not to buy resources 2. If the output of the purchase of resources is equal to the output of the non-purchasing resources, there is no spillover.

Because $\Delta Y_1 = 0$ shows that the resource trade does not generate a spillover, $\Delta Y_1 = 0$ is defined as the zero point of resource spillover. First, the type (4.4.27) is divided by${K_1}^{\eta_1+\alpha_1}({N_1}^* {K_1}^{-\psi_1})^{(1-\alpha_1)}$, and the

$$\frac{\Delta Y_1}{{K_1}^{\eta_1+\alpha_1}({N_1}^* {K_1}^{-\psi_1})^{(1-\alpha_1)}} = \left[(1-b)^{(1-\alpha_1)} - 1\right] + P({N_1}^*)^{\alpha} \frac{{K_1}^{-\psi_1\alpha_1}}{K_1^{\eta+\alpha}} b \tag{4.4.32}$$

In (4.4.32): first, $0 < b \le 1, 0 < \alpha_1 < 1$, so $\left[(1-b)^{(1-\alpha_1)} - 1\right] < 0$; second, $\psi_1\alpha_1 > 0$, so K_1 is bigger, $P({N_1}^*)^{\alpha_1}\frac{{K_1}^{-\psi_1\alpha_1}}{{K_1}^{\eta_1+\alpha_1}} b$ is smaller. Thus, it can be considered, in K_1 to a certain extent, the sale of resources is not cost-effective; of course, if the

Table 4.5 Values of the parameters of region 1

Parameter	Meaning	Initial value
b	The sale of resources accounted for the proportion of total resources b	0.1
α_1	Capital elasticity α_1	0.6
P	Resource price p	10
N_1^*	Total amount of renewable resources $N^*{}_1$	10,000
η_1	Regional technical progress factor η_1	0.3
ψ_1	Resource spillover coefficient ψ_1	0.2
K_1	Capital K_1 (can be given range)	0–50,000

P is bigger, then $P(N_1{}^*)^{\alpha_1}\frac{K_1{}^{-\psi_1\alpha_1}}{K_1{}^{\eta_1+\alpha_1}}b$ will be a large number. In addition, the increase of P promotes the sale of resources in a linear form.

To analyze the problem further, Wang and Cai simulated the two regions of the interaction process. To carry out the simulation, this paper first gives the initial value of each parameter, then we adjust the value of each parameter separately, to investigate the changes of the situation; then, approximately the resource spillover is zero, to adjust the parameters of each parameter and to investigate the changes in the parameters of the various parameters. Initial values of the parameters are shown in Table 4.5.

First, consider the impact of ψ_1 value on regional resource spillover. Because of the negative effect of capital accumulation, when capital is increasing, the amount of resources can be reduced. The greater the ψ_1, the greater the intensity of the capital for resource regeneration, the more seriously the resource decreases with the increase of capital, so it can be predicted that when the ψ increases, it will lead to the zero point of resource spillover. Figure 4.8 shows the situation of $\psi_1 = 0.1, 0.2, 0.3$; with the increase of ψ_1, the curve is more rapid than the zero point of resource spillover, and region 1 will be lost in the resource trade. That is to say, the weaker the resource regeneration ability, the more easily it can be damaged in the regional trade. Differing from general understanding, the simulation reveals that the function of the regenerative capacity is nonlinear, and the ability to improve the regeneration of the resources could be significantly improved.

Second, the influence of technology progress in the region of ΔY_1 is investigated, and the results of $\eta_1 = 0.2, 0.3, 0.4$, are obtained from the results of Fig. 4.9. Because the technology progress rate is η_1, the K_1 generated by the N multiplier effect is greater, that is to say, the sale of that part of the resources Nb, if not sold and put into production, will generate more income. Therefore, the greater the ΔY_1 will lead to the region 1 in the K_1 hours to reach the zero point of resource spillover.

Third, considering the impact of different α_1 values on the resource spillover, although the α_1 value is the capital elasticity, the α_1 value is greater than the other factors of production function. However, as in the type (4.4.32), α_1 appears in the first terms on the right. Figure 4.10 draws $\alpha_1 = 0.4, 0.8, 2$, respectively, with $P = 10$ and $P = 50$ for 0.6 cases, resource spillover with the change of K_1. Figure 4.10 shows that the location of the spillover point is not simply changed with the change

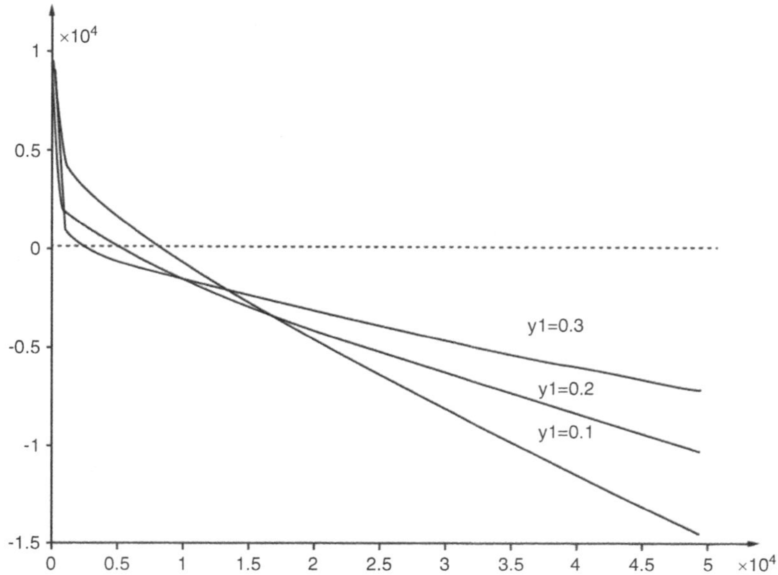

Fig. 4.8 The effect of different values on resource spillover

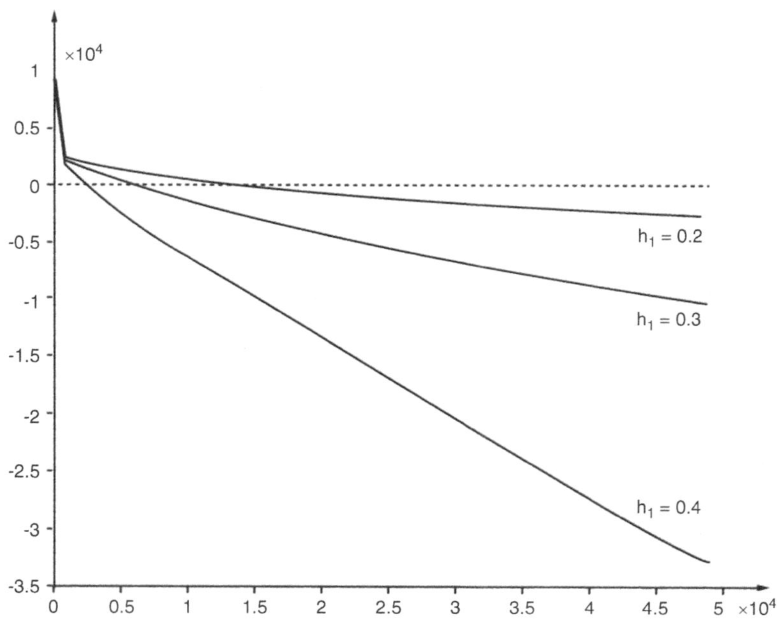

Fig. 4.9 The effect of different values on resource spillover

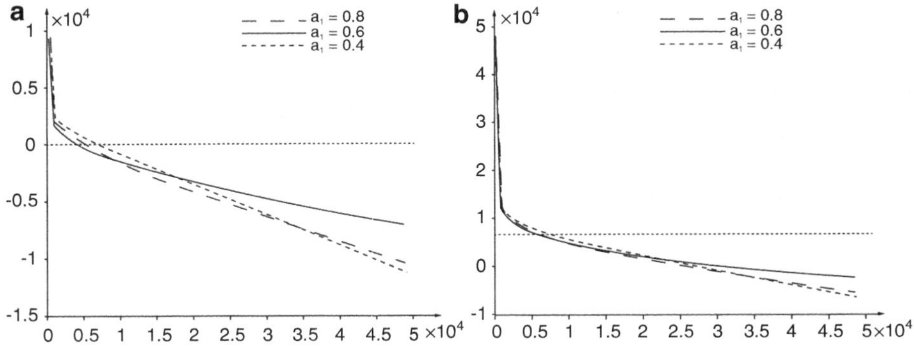

Fig. 4.10 The effect of different values on resource spillover

of α. In the picture of region 2, the K_1 of the $\alpha = 0.6$ is the smallest, in other words, the possibility of region 1 will be the lowest in the resource trade in the vicinity of $\alpha = 0.6$. By Fig. 4.10a, b, we can find the price increase to expand the scope of the capital to maintain a positive spillover of 1: the data show that the expansion of the scope of this capital is close to linear, that is, when the price increase is fivefold, the capital is also expanding in scope almost fivefold.

Similarly, the simulation also obtains resources in the resource spillover (in region 2). Then, it is possible to achieve a win–win situation by adjusting the volume of trade. Figure 4.11 shows a situation in which the two curves are represented by ΔY_1 and ΔY_2, with the change of the ratio of resource to the sale of B. O_1, O_2 are ΔY_1 and ΔY_2, respectively, by positive change to negative position. When the proportion of resources to sell is $b < O_2$, the two regions will benefit in trade; when $O_2 \leq b < O_1$, the sale of resources to region 1 is favorable, and that to region 2 is unfavorable; when $b \geq O_1$, the sale of resources on either side is not good. Therefore, if $b < O_2$, the resource trade will occur. From the picture, we can see that the maximum value of the two regions in the resource trade appears in different places; region 1 appears in the location of $b = M_1$, and region 2 appears in the position of $b = M_2$. Clearly, the two regions are always hoping to get $b < O_2$ greater spillover in the resource trade, and $b = M_1$; because of O_2, it cannot occur, and at $b < M_2$, region 2 of the spillover has not reached the maximum, and region 1 is also small, if the resource trade occurs, although it is a win–win, it is not the best. When $M_2 < b < O_2$, with the increase of b, region 1 is getting more and more large, and region 2 is reduced. Because of the existence of spillover, this range is the most likely to occur in the region of resources (see Wang et al. 2008a).

The mutual influence of multi-regional economic growth is a prominent economic problem. Through the simulation of the model, (1) in the region where technological progress is very strong, the resources produce more than their sale to obtain income. Knowledge of the region is the key to the natural resources of the region. (2) From regional decision making, the proportion of the sale of resources can be adjusted, such as by the oil output organization according to oil price. When adjusting the price of the sale of resources by adjusting the amount of resources, the

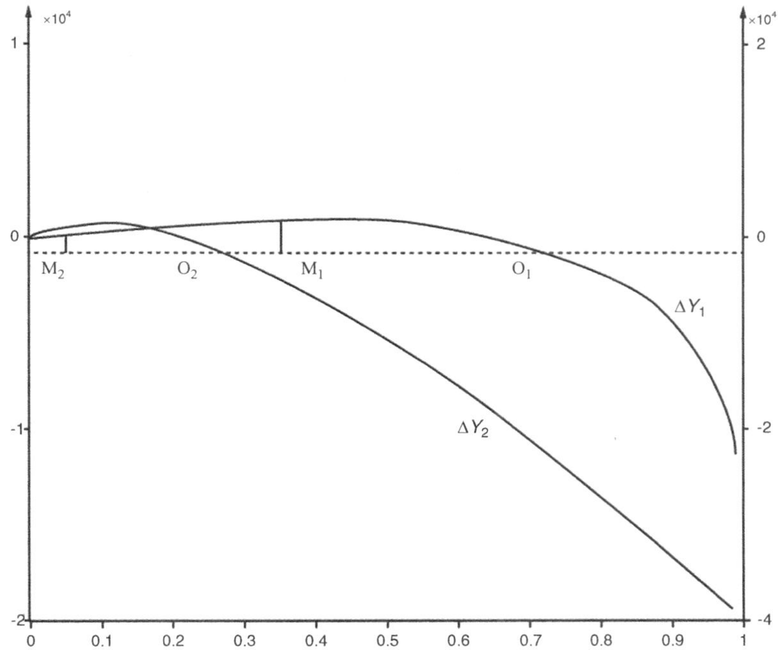

Fig. 4.11 Growth impact of the proportion of resources on the sale of two regional spillovers

region can also achieve a win–win situation. (3) The existence of a win–win situation can build a foundation for regional economic integration.

4.5 Regional Structure Evolution

4.5.1 Structure

As a real geographical phenomenon, the region is a structure. The structure of the area results from its three characteristics. The first is the whole area. Li (1959) developed a holistic view. In modern system theory, the whole of the region is characterized by a certain consistency in the region or the response characteristics of a certain geographical process, and only when they are a whole. However, in the integrity of the same area, from the perspective of observing the landscape, you can identify the river, mountains, hills, plains, and other types of settlement forms; the region and its large, small, and medium cities, towns, and villages are not homogeneous. The units in these regions need to cooperate with each other, which allows the region to form a unified function. There is a unified observation of the characteristics of the geographical matters.

Li Chunfen (1959) emphasizes that the second type of basic properties of the region are the structural characteristics, which can be referred to as the structural difference of the region. In fact, this structure makes the units cooperate with each other in the basic structure of the difference. The region contains a number of different landscape elements, landscape units in the space to form a certain structure, so that the region shows a different structure in physics and culture. In Wang et al. (2010a, b, c, d, e), the regional structure is summarized as the resource environment, population structure, urban structure, spatial structure, geographical structure, and regional economic structure (economic system); in addition, Wang and others recognize natural (ecological) zoning and ecosystem stability, in fact, the existence of ecological structure. In this book, we discuss the spatial and ecological processes and global issues, so we do not focus on the spatial structure, ecological structure, and geographical structure, and instead focus on the model of urban and rural structure and the regional economic system. Because of the complexity and the complexity of the objects, there is no complete universal model of the resource and environment structure.

Krugman (1991b) emphasized that there are two geographical natures in the region with its growth. Wang et al. (2008a, 2010a, b, c, d, e) found by research two regional geographical elements of economic difference in China: the two regional geographical natures have evolved into three natures; briefly speaking, the evolution of the region and natural endowment, traffic and gathering, information facilities, and general public service facilities. These three natures determine the structure of the region. The first nature has the leading role in an agricultural society and results in the Chinese agriculture spatial lock. In an industrial society, factors of the second nature such as transportation and infrastructure with population and industry gathered are key in the regional economics. The third nature includes the effect of human resources and regional informational and public institutions; it shows as a power in the third revolution, namely, the new economic revolution. Thus, the three natures promote regional variation. However, in this section we are limited to discussing the representation of the regional structure and avoid discussing the three natures. With the view of the three natural roles in order of evolutionary economic geography and the evolution of the early nature of historical structures, these structures sometimes produce an evolutionary lock.

4.5.2 Urban System

4.5.3 Pareto Distribution

The second nature of the population and industry gatherings, which makes the population stable when living in a certain city or town, from the first nature of the trade and the second nature of the traffic makes the cities and the towns form a system, which is the urban system. Cities and towns are different in scale, which leads to the existence of complex levels, hierarchical relationships, or connections in a system of towns, which forms the basis of the hierarchical structure of the urban system. In fact, the scale difference is very large; there are thousands of people, but also in a large city, mainly by social production mode, in the same way of production, the difference is very important. For each town, there is an impact of the size of the population on the geographical conditions of natural, economic, and social factors, but also a historical background. For example, Chinese researcher, in the mathematical analysis of the urban and town (system) in Tai'an, China, found the factors affecting the scale of the urban area, which are mainly mineral resources, basic industry, traffic conditions, and tourism resources, from a single town, show that size and development have a great relationship. However, this scale is a unique part of the regional urban system; basic industry and traffic conditions are the product of the urban system.

In theory, many scholars have studied the scale of urban system and its hierarchical relationship. In the first, Auerbach (1913) found that urban population data of five European countries and the United States were in accordance with the formula shown next:

$$P_iR_i = K \tag{4.5.1}$$

P_i is a country's urban population size, from large to small cities by population size; R_i is the first order of the city of i; K is a constant. Subsequently, many scholars have carried out the revision and growth for this theory. Zipf (1949) studied the scale distribution of the cities in developed countries and thinks that the position and scale of the city is a perfect right-angle hyperbolic relationship; this later developed into a famous city Zipf rule. Zipf (1949) believes that the size of the city is not only in line with the Pareto distribution, and the Pareto index is equal to 1, but that the corresponding A is the largest city in the urban system population. Zipf believes that the size distribution of the city is a special case of the rank-order scale rule.

The Zipf theory is this statement: if we wish urban size in the region, according to the population sorting, the rule of city size distribution can be expressed by the following formula:

$$Y = AX^{-\alpha} \tag{4.5.2}$$

In logarithmic form this is

$$\log Y = \log A - \alpha \log X \tag{4.5.3}$$

Among these, X uses population to measure the size of the city, Y is the population number of the size of the city's population, and X and A are more than 0 constant. Among them, the Pareto index of the city size distribution, the equilibrium degree of urban scale distribution in the urban system, is indicated. If a value is close to 1, the size of the city is close to that of the Zipf rule. The proportion of different cities is reasonable if the different levels of the city ply their respective functions, and all the city is in common development; if the value is more than 1, the size distribution in the city system is relatively small, with a gap between the city and another city. The higher the value, the greater the city size distribution in the urban system. All cities in the urban system have the same population size when the cities are equal. However, if a value is relatively large, the gap of all cities in the urban system is relatively small, which can cause loss of function of the urban system. If the value is less than 1, the city size distribution is relatively scattered. When = 0 is a city, there is only one city. Similarly, if the value is too small, the gap between the urban systems in all cities will be relatively large; it will cause the loss of part of the urban system.

Reasonable understanding of the urban scale meets Zipf rules. There is no theoretical proof. However, if the distribution of the city is not reasonable, but there is a city system, in the natural, economic, and social environment, this kind of existence has a certain rationality. So, be sure to be using a reasonable value. The more direct problem is that people have turned to the Pareto index, asking what factors affect the Pareto index, especially when the relationship between the Pareto index and the level of urban economy, the analysis of the impact factor of the Pareto index, has become a way to find the factors affecting urban development.

Using the data of 73 national urban populations, respectively, Soo (2005) used OLS estimation and HILL estimation of the two methods to verify the correctness of the city Zipf rules; he found, in most cases, the city size distribution of the Pareto index is not equal to 1, that is, the majority of the city's Zipf rules are not set up. He also uses the Pareto index and the potential impact of the city size distribution analysis; the results show that the political and economic factors are the main factors that affect the size of the city.

In fact, the Pareto index is a representation of the size of the city, which has a variety of interpretations. The study by Rosen and Resnick (1980) found that in the Pareto index and the per capita GDP, the total population, railway density is positively related to the land area. Mills and Becker (1986), in their study of the urban system in India, found that the Pareto index was positively related to the total population and the percentage of manufacturing workers. Alperovich (1993), using Rosen and Resnick (1980) calculated by the Pareto index, found that it is positively related to the per capita GNP, population density, and area, and to the government's share of GDP and the manufacturing industry in GDP. As the size distribution of the city is a kind of geographical feature, the influence of the political and economic factors reflects one side of the relationship between people and land. Fujita et al. (1999) believes that the larger the size of the economy, the lower the transportation cost, the smaller the proportion of the manufacturing industry in the economy, the smaller the proportion of the international trade in the economy, the more unbalanced the economic activity, and the smaller the Pareto index. When Wang et al. (2008a, b) studied the urban system in China, they found that the development of the western region does not mainly depend on agricultural production potential, population, area, and urban population ratio as much as another significant factor, the distribution of the city, and water resources, mineral conditions, and water resources. The first nature fraction is the basic driving force for urban system development.

In the world, Ettlinger (Ettlinger and Archer 1987) found the α value is gradually reduced. In the economic growth in the 1900s and 1980s, αvalues were 0.683 and 0.634, and in the war years α was 0.715. In Xu (Xueqiang) and another 100 cities in China before the scale of statistics, and the 1953, 1963, 1973, and 1978 periods of time, the size of the city of China is the size distribution ($\alpha < 1$), and α has a decreasing trend. According to this characteristic, we conclude that the α value of 0.6–0.7 may be reasonable, but it has not been proved.

In short, the complexity of the geography problem causes the complexity of α, which cannot exist as the most reasonable α value. $\alpha = 1$ is α real, but not a "reasonable" level of Zipf observation.

We need to add is that a simple analysis of the urban system in the application of geography is called the first place, which is defined as the ratio of the number of cities in the population (or GDP) to the second place in the first place, or GDP.

$$S = \frac{P_1}{P_2} \tag{4.5.4}$$

S is the first degree, P_1 is the first urban population (GDP), and P_2 is the second urban population (GDP). The significance of the first degree is that it is a simple measure of the urban system.

4.6 Hierarchical System

The structure of the urban system not only shows the statistical law, but also the hierarchical characteristics. First, we found the hierarchical nature of the structure of the central hinterland. A number of secondary cities around a large city is economic and political cooperation, and the secondary cities have a lower level of the city with the formation of a hierarchical structure. The most typical theory is the central structure of Taylor Chris, who formed K = 3, 4, 5, 7 of the space organization form. In this space, we have explained this structure from the phenomenological point of view. In the process of the region, we need to understand this phenomenon in a macro way, which requires us to understand it in the course of learning. This understanding requires the recognition of the hierarchical model of the urban system.

The starting point of the hierarchical model is not only a series, but also the economic and administrative level control over the lower-level cities.

The hierarchical (hierarchy) model can be divided into two types from different hypotheses. The Tinbergen model (Tinbergen 1961, 1968; Parr 1987) is the first. The Tinbergen model introduces the following assumptions: the urban system set contains 0, 1, 2,... N. The city level, among them, 0 rural cities; the level is set to <m>, with mth city products exported only in a 0 to m−1 level city, and other products for an export-level city similar to the other cannot be replaced; it is the same as the <m> elements of the nature of the product produced only in the mth class city, the city such as M + 1, with the nature of production elements of <m> products only for their own consumption; in each city only one product manufacturer, and the equivalent; 0 cities (rural) to provide the product to the 1 N city level; the city needs to decide on their own level of supply, to achieve a balance in the economy; the city at all levels of the level of demand for each product at a constant level, and the definition of an $a0+$ $a1+$ to meet the demand... + an + ... + $aN = 1$.

Under these conditions, the level of supply (purchasing power) of the total demand in rural areas should be equal to the supply level of rural areas (production capacity).

$$Y_0 = a_0 Y \tag{4.5.5}$$

Type Y_0 is for the supply of 0 cities (rural), Y for the total supply of all cities in the region (or total demand). For the tier 1 cities, it only exports its own products in the 0 cities, so its production capacity should be equal to the demand of the 0 cities, namely, a1Y0; at the same time it needs grade 0, 2,..., N city products, the demand for Y1 (a0 + a2 + $\cdots$ + aN) or Y1 (1−a1), then there is

$$a_1 Y_0 = Y_1(1 - a_1) \tag{4.5.6}$$

Note (4.5.5) is the available type:

$$Y_1 = \frac{a_0 a_1 Y}{1 - a_1} \tag{4.5.7}$$

For the two cities, 0, the level 1 city needs it to get the product. The production capacity can only be A_2 (Y_0 + Y_1), and it takes Y_2 (a_0 + a_3 + $\cdots$ Y_2 (+aN)− (1 − a_1 − a_2) as the demand; the result is

$$a_2(Y_0 + Y_1) = Y_2(1 - a_1 - a_2) \tag{4.5.8}$$

[Reference (4.5.6), (4.5.8) type]

$$Y_2 = \frac{a_0 a_2 Y}{(1 - a_1)(1 - a_1 - a_2)} \tag{4.5.9}$$

Completely similar to the relationship

$$Y_m = a_0 a_m Y / \left[\left(1 - \sum_{i=1}^{m-1} a_i \right) \left(1 - \sum_{i=1}^{m} a_i \right) \right] \tag{4.5.10}$$

Here we can make the economic scale of cities at all levels Y_m.

We need to further discuss the frequency of the emergence of each city. Set up an m-level city corresponding to the production capacity of the product Cm K (k), and by its supply of the city's total demand for AM (k) (Y0 + Y1+)$\cdots$ + Y_m), then the m-level cities have to be F_m:

$$f_m = \frac{a_m(k)}{C_m(k)} \sum_{i=0}^{m} Y_i \tag{4.5.11}$$

The use (4.5.11) of Yi from 0 to and 4.5.12 (m) can be obtained

$$f_m = n_m(k) a_0 \Big/ \left(1 - \sum_{i=1}^{m} a_i\right) \tag{4.5.12}$$

$n_m(k) = a_m(k)Y/C_m(k)$ by definition.

The second types of the general GH model (hierarchical) are developed by Beckmann and McPerson (1970). The GH model has been modified by the Tinbergen model, and the <m> is not only from m-level cities, but also from m + 1 to N. A <i> is a product supplied by the m-class city, and finally they prove that when the m > 0 is

$$Y = \sum_{i=1}^{m} b_{im} \sum_{h=0}^{i-1} Y_h \Big/ \left(a_0 + \sum_{i=m+1}^{N} a_i\right) \qquad m > 0 \tag{4.5.13}$$

or when $m = 0$ was established

$$Y = a_0 Y \, m = N \tag{4.5.14}$$

In the Christaller–Losch model, if the m-level cities control K_m cities, such as $K = 7$, the center of the structure control itself and the secondary six cities, then the frequency of M-grade cities is S when $m = N$:

$$\begin{aligned} f_m &= S(K_{N-1} - 1) \quad \text{when} \quad m = N - 1 \\ & S\left(\prod_{j=m}^{N-1} K_i - \prod_{j=m+1}^{N-1} K_j\right) \quad \text{Other} \end{aligned} \tag{4.5.15}$$

Establishment. Here FM is the city frequency. For the Taylor–Chris structure, K = 3, 4, 5, 7, 9, 12, or 16. A simple town system may only have a K value. A complex urban system may have many K, such as the Yangtze River Delta cities and towns system (Table 4.6), about $K = 3$, medium cities and small cities about K = 4, 7, which says that in the Yangtze River Delta cities more active in modern industry, business activities, and small cities and towns are the main link between the secondary, only $K = 4$. This scaling shows that there are binary spatial economic structures in the Yangtze River delta.

Table 4.6 Urban system in the Yangtze River Delta (1996 data)

Super-mega cities	Mega cities	Big city	Medium city	Small city	Town
1	2	4	7	22	137

4.7 Functional Structure

Cities and towns differ and thus have their own specific functions. The same city may have two or more than two functions, but not a great many.

The function of the town is not even one, but is on the urban system. Phibrick put forward the theory of central function; the center of the town functions are divided into seven or seven levels. The lowest level is as a consumer (consumer), followed by the consumption of the retail industry (retail), once again is the wholesale (wholesale), conversion (transshipment), and exchange (exchange); the top two levels are control (control) and leadership (leadership). In fact, for the modern city in wholesale, after more performance for communication, finance, trading, and technological innovation, control and leadership are achieved through these factors.

Cities and towns have the functions of these different levels, attributable to the two factors of the economy and the information economy. The gathering economy is the concentration of people's activities in the cities and towns. The result of the economy is that the enterprises have to reduce average cost. For example, in retail business, the distribution of stores can lead the stores to compete, so that consumers reduce spending, and provide the same or similar goods, so as to further attract the focus of consumer shopping activities. The wholesale and conversion is the function of the node of the input and output of the urban commodity, and it also belongs to the economy. The information economy refers to the role of the towns in the process of collecting, classifying, and spreading the information. Urban agglomeration provides the location of the accumulation of information; some large companies are headquartered in large cities, so they can widely collect information on economic activities to make decisions. The larger the size of the town, the more powerful it is to collect, transform, and spread information. In the modern economy, we can see some cities as a regional economic hub, with the role of regional "leadership," such as Shanghai, Shenzhen in China's economic status. This hub, in fact, is a functional division. From the functions of identification, the first-class hub is a research and development hub, in such a city focused on the university, research institute, knowledge service, and information service industry accounted for a large proportion in the region's lead. The second type of hub is the trade hub, their financial services, information services, logistics services, and legal services industry. The third category is the production hub; this hub of the city has a large-scale modern manufacturing base and developed logistics industry system. Hub function classification, in fact, is the result of the city through the regional spillover of the city; the city is in the spillover of modern leadership. Administrative leadership is still there, although in modern times, the importance of administrative leadership is reduced.

The existence of the hub leads to the existence of a regional hub, which is different from the center and the hinterland. A city is not the only one that belongs to each center, but is a node on the network. Given an area of professional characteristics, it is possible to identify as a trading financial hub, a number of urban nodes, mainly through the logistics and capital flow and their network. A node may be a hub in an industrial perspective, in another industry it may be ordinary nodes. For example, in China's electronics industry to Beijing, Shanghai, Shenzhen is the hub, but also has the role of high-tech equipment manufacturing industry in the hub of Shanghai, Xi'an, Shenzhen, to provide electronic, audio, and other equipment for these two. So far, it has not been seen on this kind of function theory. The theoretical model of urban functions is the need for more development. Below we speak only from the perspective of professional divisions of labor.

The function of the town is the result of specialization in the sense of economics because in the urban system, the function of the town is nonindependent. Because of the differences between the natural conditions and social economic conditions, the geographical position and the development history of the towns are different. Therefore, the functional advantages of each city are different, each city is in the competition according to the laws of the division of labor, the formation of clear urban types, such as integrated urban, industrial city, transportation hub, tourist city, etc.. Functional types and grades together constitute an aspect of the urban system. On the job scene, the urban system is not only hierarchical, but is also more strongly expressed as complementary. In the hierarchical and complementary relationship, a regional urban system has unified functional characteristics and a position in the regional and urban system. The function of the region, especially its economic function, is often been determined by the performance of the overall function of the urban system.

One way to measure the function of the city is to measure the level of the city. To measure the level of specialization of industry development, of the many which are the most commonly used is the location quotient index, that is, LQ (quotient location). LQ can reflect the relative concentration of the industry and the level of specialization, and then reflect the relative development level of the industry. The formula is

$$LQ_{ij}^{t} = \left(E_{ij}^{t}/E_{j}^{t}\right)/\left(E_{ir}^{t}/E_{r}^{t}\right) \quad (4.5.16)$$

In the form of E_{ij}^{t}, the output value of t in the j region of the i is studied, and the total output value of the E_{j}^{t} in the j region of t, E_{ir}^{t} for the large area of the study area (generally refers to the national) of the output value of j, i, t for the total output value of large area E_{r}^{t}. All the variables are in the t period. If the value of the location quotient of a sector is close to 1, it indicates that the Department's research area and the regional (i.e., the national) are similar, and the economic landscape of the Department is relatively dispersed. The location quotient of a certain area is more than 1, which indicates that the relative concentration and specialization level of the Department in the region are relatively high, the development level of this industry in the region is higher, and the development of the region has an important role. The location quotient is less than 1, then the opposite. The location quotient is the professional development level and not the scale of the industry. Obviously, if the department of the study of the distribution of the importance is low, with the perspective of regional specialization, then the value of the location should be significantly less than 1, so the region is not a corresponding industry specialization region; if the industry is important in the region, then the value of the location should be significantly greater than 1; then in the occurrence of specialization, if the industry is fully dispersed within the sub-region, the regional location should be close to 1. The existence of specialization has led to the division of functions of the urban system.

Another professional measure is the *Concise Oxford Dictionary of Geography* (Mayhew and Penny 1992), which is the industry of the two-dimensional vector space:

$$S = \left(\sum_{j=1}^{m} P_j^{\,2} \right)^{\frac{1}{2}} \tag{4.5.17}$$

In this formula, S is the specialized index and P_j is the employment population in the proportion of the industrial sector J. It has been generally considered that the S is more than 0.7 and less than 0.3. Using the employment population to calculate the GNP of the regional population, it has been determined more by the regional population. Some authors use department GNP to replace the Department's employment.

In the professional era and even earlier, the industry spatial aggregation, which accompanied the industrial regional concentration, is similar to our use of the formula to calculate the industrial regional concentration level, that is,

$$C_i = \left(\sum_{s=1}^{m} P_{i,s}^{\,2} \right)^{\frac{1}{2}} \tag{4.5.18}$$

In the formula, C_i is the aggregation index of regional i, $P_{i,s}$ is the proportion of total industrial output value of industrial output value of the city s in the region of i, which reflects the industrial concentration level within the i region. The higher the level of aggregation, the more detailed the functional division of the region.

The division of functions of the city show the need for a holistic consideration, but not just professional and gathering. Wang et al. (2004) and Lyons et al. (2001) set the LMM model to complete this analysis. An application of the LMM model is to select a city's largest number of industrial sectors, such as the basis of the production sector, 1; the need for market advertising service sector, 2A, and the need to develop the department, 2R, and both need to develop an advertising department, 2AR. These sectors are the food processing, garment manufacturing, equipment manufacturing, and electronic communications industry, which are the four production/living mutually related industries. Each industry is first calculated for professional status.

$$M_i^{(j)} = Y_i^{(j)}/Y^{(j)} \quad (4.5.19)$$

$Y_i^{(j)}$ in the i city is j industry's total output value, $Y^{(j)}$ for the entire region is j industry's total output value. Obviously, this value could be calculated by the location quotient. The selection of the four departments or several departments reflects the function of the urban system. The LMM model is further improved, and the empirical formula has been established by using a complete economic system (such as the national).

$$S_i = \beta_1 M_i^{(1)} + \beta_2 M_i^{(2A)} + \beta_3 M_i^{(2R)} + \beta_4 M_i^{(2AR)} + \mu_i \quad (4.5.20)$$

Further, we need to calculate the industry market. In the case of economic crisis, the market supply and demand balance, the market share of an industry in a region, is equal to the output value of the industry. The market share of j industry in i region can be estimated by the proportion of total output value of j industry in j city.

4.7.1 Regional Economic System

Based on the three nature effects, the basic structure of the region is that it is an economic system. Economics is about the economic system of the model, but from theoretical geography, we can suggest that the regional economic system involves three levels of economic problems: the first is the regional macroeconomic framework, the second is the microeconomic expression, and the third is in the region of economic innovation, a regional economic system model; it should be described in the three levels of economic system. A number of representative models can start this discussion. Of course, in the actual problem analysis, we can cut into the regional economic issues from these three levels.

4.8 Macroeconomic Framework

Lorentz and Savona (2008) constructed an industrial evolution model based on the input–output model. As a model of modern economics, it needs to have the function of the industrial evolution process, which is driven by innovation. Such a model could be divided into two major sectors, including the macro level of economic structure evolution and the technological level of enterprise behavior change, which will affect macroeconomic growth and structural change.

Leontief (1951) proposed the input–output framework as suitable for the analysis of macroeconomic issues, and then the input–output model $Y_{j,t}$ of the input–output model will be divided into three parts: the middle consumer $I_{j,t}$, domestic final product consumption $C_{j,t}$, and net foreign consumption $(X_{j,t} - M_{j,t})$. Total output is a function of the sector structure of the economy, which is determined by the demand for intermediate goods and final products:

$$\begin{pmatrix} Y_{1,t} \\ \vdots \\ Y_{j,t} \\ \vdots \\ Y_{J,t} \end{pmatrix} = \begin{pmatrix} I_{1,t} \\ \vdots \\ I_{j,t} \\ \vdots \\ I_{J,t} \end{pmatrix} + \begin{pmatrix} C_{1,t} \\ \vdots \\ C_{j,t} \\ \vdots \\ C_{J,t} \end{pmatrix} + \begin{pmatrix} X_{1,t} \\ \vdots \\ X_{j,t} \\ \vdots \\ X_{J,t} \end{pmatrix} - \begin{pmatrix} M_{1,t} \\ \vdots \\ M_{j,t} \\ \vdots \\ M_{J,t} \end{pmatrix} \tag{4.5.21}$$

The intermediate product department j $I_{j,t}$ is the sum of the j product requirements of various departments, defined as

$$I_{j,t} = \sum_{k=1}^{J} Y_{j,k,t}^{D} = \sum_{k=1}^{J} a_{j,k,t} Y_{k,t} \tag{4.5.22}$$

where $Y_{j,k,t}^{D}$ is the demand for the j department of the k sector, $Y_{k,t}$ is the output level of the k sector, the intermediate demand coefficient $a_{j,k,t}$:

$$a_{j,k,t} = \sum_{i} z_{k,i,t} a_{j,k,i,t} \tag{4.5.23}$$

in which $z_{k,i,t}$ is the k sector i enterprises accounted for by the market share, by the (4.5.23) type decision, $a_{j,k,i,t}$ is the i sector k sector j department of intermediate product consumption coefficients. I_t (4.5.22) in the intermediate product consumption matrix form is expressed as

$$\begin{pmatrix} I_{1,t} \\ \vdots \\ I_{j,t} \\ \vdots \\ I_{J,t} \end{pmatrix} = \begin{pmatrix} a_{1,1,t} & \cdots & a_{1,k,t} & \cdots & a_{1,J,t} \\ \vdots & \ddots & \vdots & & \vdots \\ a_{j,1,t} & \cdots & a_{j,k,t} & \cdots & a_{j,J,t} \\ \vdots & & \vdots & \ddots & \vdots \\ a_{J,1,t} & \cdots & a_{J,k,t} & \cdots & a_{J,J,t} \end{pmatrix} \begin{pmatrix} Y_{1,t} \\ \vdots \\ Y_{j,t} \\ \vdots \\ Y_{J,t} \end{pmatrix} \tag{4.5.24}$$

Final consumption $C_{j,t}$ is a function of the total real income level. Assume that the final consumption of the j sector accounted for the proportion of total real income levels of $c_{j,t}$, then:

$$C_{j,t} = c_{j,t} \widetilde{Y}_t \tag{4.5.25}$$

Among them, the actual income level of $\tilde{Y}_t$ and the actual related GDP, is the nominal output of the department in accordance with the sum of the total price level, so the final consumer $C_{j,t}$ could be expressed as

$$C_{j,t} = c_{j,t}\sum_{k=1}^{J}\frac{p_{k,t}}{\bar{p}_{t-1}}Y_t \tag{4.5.26}$$

where $\bar{p}_{t-1}$ is total price level, $p_{k,t}$ is the sector price level, and $Y_t$is the nominal output. The final consumption of C_t can be expressed in a matrix form:

$$C_t = \begin{pmatrix} C_{1,t} \\ \vdots \\ C_{j,t} \\ \vdots \\ C_{J,t} \end{pmatrix} = \begin{pmatrix} c_{1,t}\frac{p_{1,t}}{\bar{p}_{t-1}} & \cdots & c_{1,t}\frac{p_{k,t}}{\bar{p}_{t-1}} & \cdots & c_{1,t}\frac{p_{J,t}}{\bar{p}_{t-1}} \\ \vdots & \ddots & \vdots & & \vdots \\ c_{J,t}\frac{p_{1,t}}{\bar{p}_{t-1}} & \cdots & c_{J,t}\frac{p_{k,t}}{\bar{p}_{t-1}} & \cdots & c_{J,t}\frac{p_{J,t}}{\bar{p}_{t-1}} \\ \vdots & & \vdots & \ddots & \vdots \\ c_{J,t}\frac{p_{1,t}}{\bar{p}_{t-1}} & \cdots & c_{J,t}\frac{p_{k,t}}{\bar{p}_{t-1}} & \cdots & c_{J,t}\frac{p_{J,t}}{\bar{p}_{t-1}} \end{pmatrix}\begin{pmatrix} Y_{1,t} \\ \vdots \\ Y_{j,t} \\ \vdots \\ Y_{J,t} \end{pmatrix} \tag{4.5.27}$$

It is assumed that the proportion of the total domestic demand of each department is $m_{j,t}$, and the net exports of the department can be expressed as $(I_{j,t}+C_{j,t})$:

$$X_{j,t} - M_{j,t} = X_{j,t} - m_{j,t}\left(I_{j,t} + C_{j,t}\right) \tag{4.5.28}$$

Equation (4.5.24) and formula (4.5.26) can be defined as net exports of $(X_{j,t} - M_{j,t})$

$$\begin{aligned} X_t - M_t \equiv & \begin{pmatrix} X_{1,t} \\ \vdots \\ X_{J,t} \end{pmatrix} \\ & - \begin{pmatrix} m_{1,t}\left(a_{1,1,t} + c_{1,t}\frac{p_{1,t}}{\bar{p}_{t-1}}\right) & \cdots & m_{1,t}\left(a_{1,J,t} + c_{1,t}\frac{p_{J,t}}{\bar{p}_{t-1}}\right) \\ \vdots & \ddots & \vdots \\ m_{J,t}\left(a_{J,1,t} + c_{J,t}\frac{p_{1,t}}{\bar{p}_{t-1}}\right) & \cdots & m_{J,t}\left(a_{J,J,t} + c_{J,t}\frac{p_{J,t}}{\bar{p}_{t-1}}\right) \end{pmatrix} \\ & \times \begin{pmatrix} Y_{1,t} \\ \vdots \\ Y_{J,t} \end{pmatrix} \end{aligned} \tag{4.5.29}$$

Equations (4.5.4), (4.5.6), and (4.5.9) as brought into the formula (4.5.1) are available:

$$\begin{pmatrix} Y_{1,t} \\ \vdots \\ Y_{1,t} \\ \vdots \\ Y_{1,t} \end{pmatrix} = \begin{pmatrix} a_{1,1,t} & \cdots & a_{1,k,t} & \cdots & a_{1,J,t} \\ \vdots & \ddots & \vdots & & \vdots \\ a_{j,1,t} & \cdots & a_{j,k,t} & \cdots & a_{j,J,t} \\ \vdots & & \vdots & \ddots & \vdots \\ a_{J,1,t} & \cdots & a_{J,k,t} & \cdots & a_{j,j,t} \end{pmatrix} \begin{pmatrix} Y_{1,t} \\ \vdots \\ Y_{1,t} \\ \vdots \\ Y_{1,t} \end{pmatrix} + \begin{pmatrix} X_{1,t} \\ \vdots \\ X_{1,t} \\ \vdots \\ X_{1,t} \end{pmatrix} \tag{4.5.30}$$

Among them

$$\alpha_{j,k,t} = \left(1 - m_{j,t}\right)\left(a_{j,k,t} + c_{j,t}\frac{p_{k,t}}{p_{t-1}}\right) \tag{4.5.31}$$

Based on the formula (4.5.30), the simplified formula of the model can be obtained under the condition of short-term macroeconomic identity:

$$\begin{pmatrix} Y_{1,t} \\ \vdots \\ Y_{j,t} \\ \vdots \\ Y_{J,t} \end{pmatrix} = \begin{pmatrix} 1 - a_{1,1,t} & \cdots & -a_{1,k,t} & \cdots & -a_{1,J,t} \\ \vdots & \ddots & \vdots & & \vdots \\ -a_{j,1,t} & \cdots & 1 - a_{j,k,t} & \cdots & -a_{j,J,t} \\ \vdots & & \vdots & \ddots & \vdots \\ -a_{J,1,t} & \cdots & -a_{J,k,t} & \cdots & 1 - a_{j,j,t} \end{pmatrix}^{-1} \begin{pmatrix} X_{1,t} \\ \vdots \\ X_{k,t} \\ \vdots \\ X_{J,t} \end{pmatrix} \tag{4.5.32}$$

Thus, in the short term, the solution is only a function of the demand and the function of the outlet. At the same time, the assumption that the department has given to the external is that the parameter is proportional to the growth. Equation (4.5.32) is a description of the internal links of the regional industry in the macro economy.

4.9 Enterprise Economic Activities

Regional economic systems often need to describe the enterprise level. In this model, its share of the demand in the sector determines the output of the enterprise. Enterprise production factors include the combination of intermediate goods and labor, Lorentz and Savona (2008) is obtained from all departments, and the production function of the enterprise has the Leontief form:

$$Y_{k,i,t} = \min\left(\frac{1}{a_{1,k,i,t}}Y^{(D)}_{1,k,i,t}, \ldots, \frac{1}{a_{j,k,i,t}}Y^{(D)}_{j,k,i,t}, \cdots \frac{1}{a_{J,k,i,t}}Y^{(D)}_{J,k,i,t}, A_{k,i,t}L_{k,i,t}\right) \tag{4.5.33}$$

where $A_{k,i,t}$ states i labor productivity. The output level of i enterprises in the k sector is the share of the total demand of the k sector $z_{k,i,t}$:

$$Y_{k,i,t} = z_{k,i,t} Y_{k,t} \tag{4.5.34}$$

Enterprise i to the j department's product demand is defined as

$$Y^{(D)}_{j,k,i,t} = a_{j,k,i,t} Y_{k,i,t} \tag{4.5.35}$$

By (4.5.33) i enterprise labor demand is:

$$L_{k,i,t} = \frac{Y_{k,i,t}}{A_{k,i,t}} \tag{4.5.36}$$

Enterprise i's market share is defined as the replicator dynamics (replicator dynamic; dynamic change of market share here is actually the process of market selection of a simple abstract:

$$z_{k,i,t} = z_{k,i,t-1}\left(1 + \phi\left(\frac{E_{k,i,t}}{E_{k,t}} - 1\right)\right) \tag{4.5.37}$$

in which $E_{k,i,t}$ and $E_{k,t}$, respectively, are the competitiveness of i enterprises and the average competitiveness of the k sector. Each enterprise's competitiveness level is defined as the reciprocal of the price level of the enterprise:

$$E_{k,i,t} = \frac{1}{p_{k,i,t}} \tag{4.5.38}$$

The price level of the enterprise is set to be added to the unit production cost $\kappa_{k,i,t}$ based on $\mu_{k,i}$, which is defined as the cost of production per unit of $\kappa_{k,i,t}$:

$$\kappa_{k,i,t} = \sum_{j=1}^{J} a_{j,k,i,t} p_{j,t} + \frac{w_{k,t}}{A_{k,i,t}} \tag{4.5.39}$$

$w_{k,t}$ is the t phase of the k sector wages. The enterprise price setting is as follows:

$$p_{k,i,t} = (1 + \mu_{k,i})\left(\sum_{j=1}^{J} a_{j,k,i,t} p_{j,t} + \frac{w_{k,t}}{A_{k,i,t}}\right) \tag{4.5.40}$$

where $p_{j,t}$ represents the average level of the j sector:

$$p_{j,t} = \sum_{i} z_{k,i,t} p_{j,i,t} \tag{4.5.41}$$

Corporate wages are set at the departmental level. For a given department k, the dynamic equation of wages relates to the labor productivity of the department k $\frac{\Delta A_{k,t}}{A_{k,t-1}}$. The department wage change is defined as

$$\frac{\Delta \omega_{k,t}}{\omega_{k,t-1}} = \rho \frac{\Delta A_{k,t}}{A_{k,t-1}} \tag{4.5.42}$$

where ρ is the sensitivity of wages to labor productivity changes. The enterprise sells the product to obtain the profit, the profit margin for the total revenue:

$$\pi_{k,i,t} = (P_{k,i,t} - \kappa_{k,i,t})Y_{k,i,t} \tag{4.5.43}$$

On the economic analysis of individual enterprises, the enterprises in the pursuit of profit (Eq. (4.5.43)), are the largest in the microeconomic sense. In the regional economic system, the regional assumption is that the enterprise will sell products and services generated by the profits of the investment market for reasonable innovations.

4.10 Technological Innovation

An important feature of the regional economy is that it has a new behavior, which has a great impact on industrial structure through total factor productivity and technological innovation. A large number of scholars have done many types of research on this innovative economics (Green and Scotchmer 1995; Lorentz and Savona 2008). In the model, the technical impact caused by technological innovation increase, and the structural change of intermediate consumption coefficient, are expressed. Because of the uncertainty of the result of innovation, it assumes that the possibility of innovation is the function of the innovation investment. Enterprises will be filled with profit for innovative investment, which is used for the share of labor innovation, for the share of material capital innovation is a. The possibility for the success of labor innovation investment:

$$\ell_{k,i,t} = {}^{k,i,tk,i,t} 1 - e^{-\alpha(1-\eta)\pi_{k,i,t}} \tag{4.5.44}$$

The possibility for the success of the investment of material capital innovation is

$$\chi_{k,i,t} = {}^{k,i,tk,i,t} 1 - e^{-\alpha(1-\eta)\pi_{k,i,t}} \tag{4.5.45}$$

If the innovation is successful, then the labor productivity $A_{k,i,t}$ and the intermediate product demand coefficient $a_{j,k,i,t}$ of the change will be a random process. In the model of Lorentz and Savona (Lorentz and Savona 2008), impact obeys normal distribution and has a direct effect on the production process, which is embodied in the absolute fluctuation of $A_{k,i,t}$ and $a_{j,k,i,t}$ in labor productivity. This approach ignores the differences between different enterprises, industries, and the level of technology. The enterprise of innovation must be based on the original human, material, technology; innovation is the result of continuous, cumulative R & D behavior. Green and Scotchmer (1995) pointed out that the knowledge and technology progress is cumulative; product is a number of step inventions (invention), modified (modifications), and improved (improvement) results, and the sequence of the process of innovation. According to the theory of sequence innovation, the innovation is based on the sequence, and the development of technology is the process of gradual evolution. This innovation is not only reflected in the new generation of products based on the development of the past generation, but is also reflected in the production process of enterprises that will continue to improve the production process to improve the production technology level and reduce production costs.

We learn from the idea of innovation sequence, and continue to use Green and Scotchmer (1995); the practice of that next level of technology change and the technical level-related innovation is caused by impact on labor productivity $A_{k,i,t}$ and intermediate goods demand coefficient $a_{j,k,i,t}$ changes rate, then the new labor productivity equations are

$$\begin{aligned} A_{k,i,t} &= A_{k,i,t-1}(1+\max\{\varpi_{k,i,t};0\}) \\ &\varpi_{k,i,t}\tilde{}N(0;\sigma) \end{aligned} \tag{4.5.46}$$

Similarly, the new intermediate product demand coefficient is

$$a'_{j,k,i,t} = a_{j,k,i,t-1}\left(1+\varepsilon_{j,k,i,t}\right) \quad \varepsilon_{j,k,i,t}\tilde{}N(0;\xi) \tag{4.5.47}$$

When the impact of the unit output is below the actual unit of $\kappa_{k,i,t}$, the output of the new coefficient set $\left(a'_{1,k,i,t},\ldots,a'_{j,k,i,t},\ldots,a'_{J,k,i,t}\right)$ is the introduction of the model, that is, the impact of the new model.

$$\left(a_{1,k,i,t+1},\ldots,a_{J,k,i,t+1}\right) = \begin{cases} \left(a'_{1,k,i,t},\ldots,a'_{j,k,i,t},\ldots,a'_{J,k,i,t}\right) & \text{if } \sum_{j=1}^{J} a'_{j,k,i,t}p_{j,t} < \kappa_{k,i,t} \\ \left(a'_{1,k,i,t},\ldots,a'_{j,k,i,t},\ldots,a'_{J,k,i,t}\right) & \text{otherwise} \end{cases} \tag{4.5.48}$$

According to Gong Yi, the process innovation of enterprise success lies in its investment that could reach a certain threshold of ζ_c innovation. The Gong et al. (2013) simulation shows that regional focus in replacement of innovation and process improvement of innovation and product upgrading describes the different effects of the Hoffman ratio of regional industrial structure (that is, the ratio of light industry and heavy industry total output value). With the increase of the scale of the process innovation, the proportion of Hoffman is on the rise. With the increase in the size of the product innovation, the proportion of Hoffman also showed a downward trend.

It is worth noting that the formula (4.5.48) is an evolutionary choice. The improvement of enterprise technology is realized through the interests of enterprises, and through this kind of individual choice, the region will eventually complete the whole technology level, which means that the new technology of the complex system of regional innovation completes the regional evolution.

4.11 Area CGE

A state of regional economy is in general equilibrium. Analysis of regional economy in the vicinity of the general equilibrium point of fluctuations is an effective analysis of the region under the impact of policies, environment, etc. In this analysis, the computable general equilibrium model (CGE model) is based on the general equilibrium theory and the economic model of a specific economic system.[7] The CGE model is a set of equations that describe the supply, demand, and market economic relationships to solve economic states, and with conditions of the producer's profit maximization, consumer utility maximization of a series of optimization constraints that the equation group in each market has reached equilibrium when a set of quantity and price reflects the economy through the adjustment of the quantity and price of goods and factors, to realize the balance of the Walrasian general equilibrium theory to describe supply and demand. The CGE model structure is divided into four major modules: product supply, product demand, revenue, and market equilibrium. The model object has enterprises, central government and local government, urban residents, and rural residents. For different regions, domestic and foreign trade has increased the area of two foreign accounts.

The CGE model was initially used to analyze the impact of policy and economic changes in the country, but with the development of CGE research, a prominent problem is how to apply the CGE model to the regional level. The reason why we need to study the regional level of the CGE model is because with the country as a whole as a research object, analysis of some problems is too general. Especially for a large country such as China, the land area is large, and each domestic region has a large difference in the resource endowment and economic development conditions. As a result of the CGE simulation of the national level, it cannot explain and guide the development of the various regions, but only in the macro level of the country. At present, the research of a multi-regional economic model has attracted much attention from the international academic field. The CGE model of multi-regions and the analysis of policy simulation are important to improve the level of government decision making and enhance the ability of the development plan.

---

[7]For more detailed introduction of the CGE model, the CGE technology of the economic development policy simulation analysis, such as Wang Zheng and Xue Junbo, is introduced.

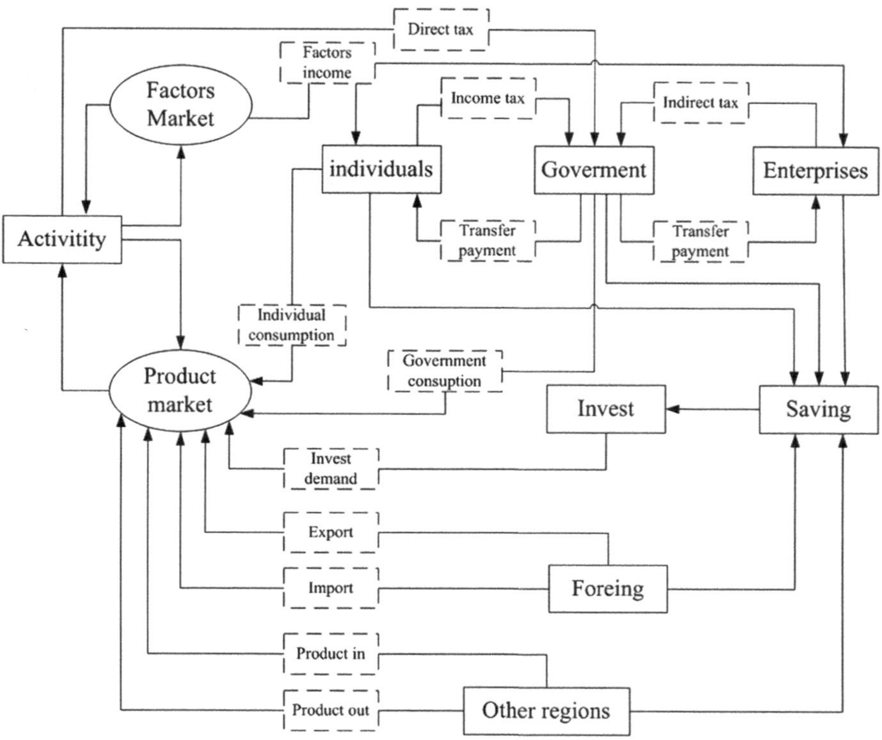

Fig. 4.12 The structure of a region in a multi-area CGE (Modified from Kim 2002)

Here we give a detailed description of the basic structure of the CGE model of the Chinese economy. It is consistent with the general CGE model, although limited as to space: this book does not discuss the specific model. For the common logical structure of the common CGE models and the solution, see Chap. 7 (Fig. 4.12).

In the CGE model, there are several noteworthy issues:

1. In a CGE model, which has a certain closeness, it is assumed that each of the manufacturing sectors in the economic system has a competitive enterprise, and each enterprise produces a commodity or service. Each production department, through the use of composite products (import goods and products in accordance with the Armington hypothesis, the use of CES transformation), the input, labor, and capital and other factors, has the production of goods for the region's consumption, domestic consumption, or export of goods or services.
2. In the process of production of CGE, the production department is not the price decision maker but the price receiver; therefore, the enterprise (department) must be in certain technical conditions, in accordance with the established cost of profit or the minimum principle to produce the profit cost. Decision making is the production possibility frontier, according to the principle of maximizing the income of the department of production for domestic and export of the relative

share of the composition. In short, under the assumption of scale, the total output of the departments cannot be determined by the producer, but by the equilibrium conditions. The producer needs to carry out the decision, in the condition of the total equilibrium conditions of the department, of the choice of the intermediate input and the effective input level, so that the production cost is minimized.

3. World modeling has a choice for the form of the production function of various departments. A practical application of the CGE model as used in the form of equations is not arbitrary, but is often selected from a number of specific equations, which are mainly related to the form of production function and consumption function. The choice of function form depends on the rationality of theory and the rationality of the analysis. On the one hand, the CGE model in the form of function but also to meet the general balance constraints, such as market clearing; on the other hand, in the important parameters of the model, function of convenience often compared to appropriate more easily determined equations of the form.

The regional economic system of the CGE and the national CGE system is the difference between the regional economy and the openness of the coexistence, in particular, of a multi-regional system; there are mutual transfer payments and trade. Interested readers might wish to see Sun et al. (2009); however, this book is no longer in print.

Chapter 5
Regional Development

Abstract Evolution, as an academic term, was initially proposed by Herbert Spencer in 1857, and he proposed that the mechanism from simple to complex is called "evolution", a term that was later introduced into biology to describe the development of life irreversibility, inheritance, and path dependence. Regional dynamics processes and spatial dynamics processes lead to regional population aggregation, road construction, environmental change, and resource use, with complete evolutionary characteristics.

Regional evolution means that in regional complex development, the driving force for growth is growth, the growth of geography is innovation, and innovation is the driving force of formal evolution. Therefore, in this book, we sometimes talk about regional development, in fact, about its evolution. In this chapter, the author discusses the regional evolution of China with the understanding of regional dynamics. Some of its underlying models and concepts are described in the previous two chapters, and the focus of this chapter is to show regional evolution through examples.

As the result of the evolution, the authors suggest that the existence of spatial organization has evolved into two kinds of display, the center–hinterland structure and the hub–network structure.

Regional economic growth, or regional growth, is the most prominent problem in the region. The need for understanding and regulation of regional growth patterns is of wide concern. This book argues that regional growth is a function of the inter-organizational population, resources, environment, and capital flow, and that capital is not only material capital, but also knowledge capital. This kind of capital flow can be within a region, and it can be between regions. Innovation activity is the activity parameter of this function, because for the same population, resources, environment, and capital under different innovation mechanisms, the growth result is not the same. As Inpet knew, innovation is the driving force of economic growth.

From a geographical point of view, one of the characteristics of the region is that a spatial network exists, so we should consider this network background from the aspect of theoretical geography.

Regional growth is achieved through regional capital, technological diffusion, and capital transfer. This process tends to have jumps. It is difficult to simulate the differential equation model using the space background. In the "space process," we

Z. Wang, *Principles of Regional Science*, New Frontiers in Regional Science: Asian Perspectives 15, DOI 10.1007/978-981-10-5367-2_5

know that the modeling base is agent based, which can depict our problems. This method, at the micro level, has the characteristics of autonomy, sociality, activity, reactivity, etc., to decompose a complex problem into a number of agents that constitute the main body of the same environment.

5.1 Three Geographical Natures and Regional Background

The three geographical natures of the region are the basis of regional evolution. These bases are also the growth of the background space. In this analysis, we assume that some places in this space can be hosted by agent-based movement (ABM). From the geographical point of view, the activities of these capital organizations are restricted by the three natures or are popularly said to be subject to geographical location and environmental conditions. In fact, the main performance of the region, and the development needs of the second and the third nature, are the regional endowment of market resources, transportation resources, social resources, and information resources. The second nature is the development of enterprises and public service resources, only possible with the corresponding sound infrastructure, to facilitate the transport of raw materials and products as well as the exchange of information and technology. These factors are also applied to the ABM study area, which needs to be considered.

Of course, the last need is to describe regional human resources. Obviously, these resources can be described as the space of the region. Wu et al. (2008) describe the resource and environment characteristics of space using the automaton model, which is based on nonuniform spatial characteristics. We can also calculate the nature of each place more generally. Figure 5.1 is a synthetic figure of the three natures as described by Wang Zheng et al. based on their computational methods (Liu and Wang 2009; Xia and Wang 2012).

5.2 Regional Evolution

5.2.1 Regional Evolution ABS

Wang et al. (2014a, b) and Gu and Wang (2014) developed a model using agent-based simulation (ABS) to describe regional growth. In this model, the authors assume that the economic system contains a number of local nodes, which represent the inhabitants of the land, towns, and cities. Their geographical nature is described in 5.2.1; we might say that they are hard and remote. In addition to geographical nature, in economic activities, these local nodes have similar properties and behavior. The term Agent represents the spatial nodes; according to the different locations of Agent, the nodes can be divided into several groups, so the economic system can be abstracted as a complex system consisting of multiple groups of Agents and the

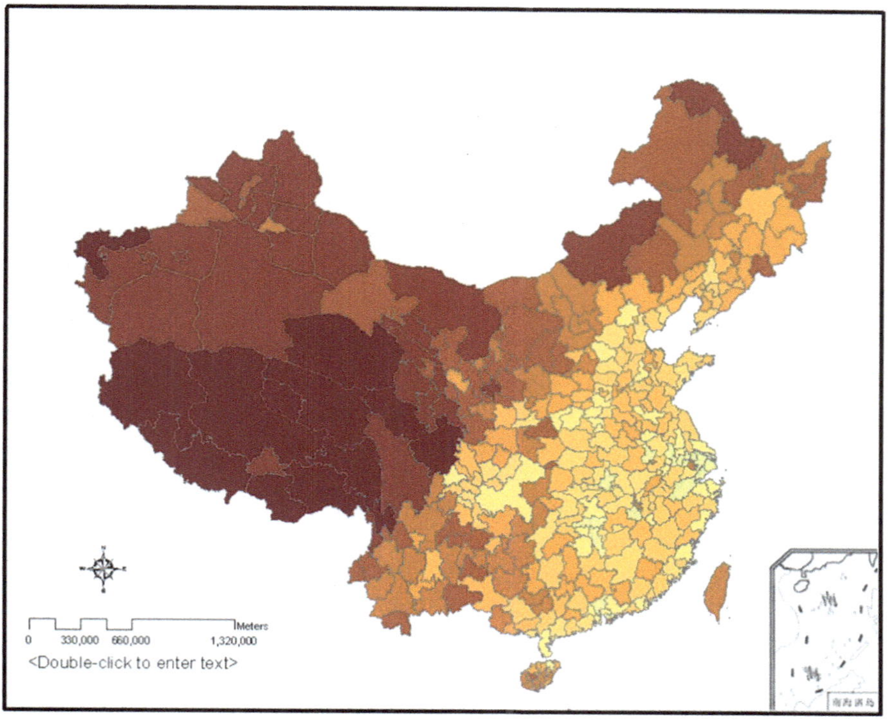

Fig. 5.1 Example: China's three natures of the right

external market environment. In a practical example, Gu and Wang (2014) showed China's 362 prefecture-level cities as autonomous prefectures, as well as by the province directly under the jurisdiction of the county defined as Agent; that is, each place has a capital organization Agent that is representative of this place, which is equivalent to the existence of a system of 362 capital organizations. In such a model, each independent body with time to change its spatial location or other geographical characteristics is the formation of a dynamic system. The spatial dynamics model (Wang et al. 2014a, b; Gu and Wang 2014) uses the Agent-based state sequence simulation; the 0th stage is the initial state of the system, which constitutes the starting point of the model. This approach is based on the main idea of building a spatial analysis model.

5.2.1.1 Hypothesis and Basic Rule Design

For the main body of the model, some assumptions and basic rules about the Agent-based movement are first required.

The established capital organization Agents in the 0th phase are all functioning, with these further hypotheses. (1) The labor force compared to the total market capital organization, Agent wages, and the degree of population congestion determines whether labor migration will allow production technology to proliferate. (2) The capital return ratio of Agent and the size of the market are determined by the capital market. (3) The net profit value of the Agent in the first period of the capital organization is the additional investment of the current period. (4) The flow and transfer of the labor force and the capital along the Agent are (5) to link the traffic road network to the local area. The distance between the grounds is an important factor that affects the interaction between the two places, and the interaction between the two meets the Wilson potential.

The difficulty in calculating the spatial interaction is to estimate the spatial damping parameter or β; the spatial damping of the local population migration is $\beta = \sqrt{\frac{2T}{t_{\max} D}}$, and the T is the average time of migration. In this case, in a case of the same domain size, the two migration intervals can reduce the spatial damping coefficient of population migration. This spatial transfer mode also applies to the transfer and flow of capital along the road transportation network.

In the real world, between regions there are many different traffic roads, such as railways, electrified railways, highways, national highways, and provincial roads. These roads have different road weights because of their different speeds and carrying capacity. To simplify the problem, set an area between transportation by the railway network, and the railway is divided into ordinary railway, electrified railway, Gao Tiesan; for no railway through the area, Gu Gao Xiang, set to the national or provincial mode instead (Wang et al. 2014b). Therefore, the road transport network used by the model is made up of four kinds of weights, and the weight of the road transportation network is $w_i = \frac{\max(t_i)}{t_i}$ $i \in [1, 2, 3, 4]$. t_i is the average time to spend on a section of an i type. Accordingly, we obtain the weight of the path of the region j to the region i:

$$W_{i,j} = \sum_{k,l} D_{k,l} w_{k.l} \Big/ \sum_{k,l} D_{k,l} \tag{5.2.1}$$

$W_{i,j}$ is the weight of the shortest path in the i region to the j region, and the $D_{k,l}$ is the shortest distance between the k l and the j to the i region, and $w_{k.l}$ is the weight of the k to the l line. In this way, the spatial damping coefficient of the i region to the j region is

$$\beta_{i,j} = \sqrt{\frac{2T}{t_{\max} D \cdot W_{i,j}}} \tag{5.2.2}$$

5.2.1.2 Agent Behavior Design

Gu Gaoxiang (Wang et al. 2014b) will be the 362 prefecture-level cities in China abstracted as capital organization Agents. In the economic world, the main activity of an Agent can be summarized as production, sale, profit, attracted investment, and labor, which is the only way to defeat competitors and survive in the economic world. Agent behavior includes the following factors.

Production Behavior Design

For the capital organization Agent, the model is defined in the form of Cobb–Douglas.

$$Q = h^{\alpha} L^{\beta} K^{1-\beta} \tag{5.2.3}$$

Q is the output of capital organization, K is the capital stock of capital organization, L is the labor force of capital organization, h is the production technology level of capital organization, α is the technical elasticity index, and β is labor substitution elasticity index. Among them, the product technology level H by the capital organization is the R & D innovation capability index decision.

$$h = \mathrm{Log}_a(he + b) \tag{5.2.4}$$

In Eq. (5.2.4), he is the capital organization Agent R & D innovation capability index, a, and b is a parameter.

Enterprises in the production process consume a certain amount of resources, such as intermediate input; this chapter in the production process of resource consumption is defined as the production process of fixed costs. The fixed cost of enterprise production to consumption for each is

$$C_p = c \cdot Q \cdot P_{R,t} \tag{5.2.5}$$

In Eq. (5.2.5), c is the proportion of the demand for resources, the market price of resources for P_R.

In the current period, the wage rate w_t of the enterprise is determined by the marginal product of labo, and the price of the product sold in this period.

$$w_t = d_{L,t} \cdot P_t = \beta h^{\alpha} L^{\beta-1} K^{1-\beta} \cdot P_t \tag{5.2.6}$$

In Eq. (5.2.6), $d_{L,t-1}$ for enterprise i in the t period of labor marginal output.

Profit from capital organization:

$$\pi = \left(M - \delta K \cdot P_{K,t} - W - C_p - C_T\right) \cdot (1 - tax) \tag{5.2.7}$$

In Eq. (5.2.7) M is the sales of enterprise sales revenue, C_T is the cost of transportation cost of goods sold, δK is the depreciation of capital, C_p is the current production of fixed costs, and P_K is the market price of fixed capital.

Labor Flow and Technology Diffusion

The model agrees that the labor force is able to flow and migrate along the road transportation network between the capital organizations, the Agents, which is in line with the actual construction. The driving labor force in the flow of Agent is influenced by many factors, which include wage difference, the degree of crowding, the distance between the hard spaces, and the Agent. This driving force is a force of space. In the course of the flow, labor tends to migrate to a city with a higher wage rate than the current city, although the population is not very dense, with less competitive pressure in the industry. The gap of wage rate is regarded as the pulling force of labor, and the intensity of population is regarded as the thrust of labor force flow. At the same time, the labor flow is affected by the distance between Agent and remote and difficult locations. The hard remote index is a kind of evaluation that is based on the natural environment and the remote nature of the human environment. The evaluation content includes altitude, surface roughness, agricultural production potential, climate condition, water resources, transportation convenience, communication facilities, and cultural differences. Based on the self-subject technique, it is especially suitable for the spatial analysis of this kind of space. The figure given in this section shows the evaluation results of remote China. We can find that China's important geographical boundaries, such as the Hu Huanyong line, and the Huaihe–Qinling Mountains line, are clearly reflected in the map.

The flow of population and capital is driven by the interaction between the space and the background of a difficult and remote nature. In the model of spatial interaction, the Wilson model is introduced to describe the attraction of Agent to Agent:

$$\begin{aligned} TP_{ij} &= (P_i \cdot \exp(ef_i))\left(w_j/\exp\left(ef_j\right)\right)\exp(-\eta D_{ij}) \\ &= AP_i w_j \exp\left(\left(ef_i - ef_j\right) - \eta D_{ij}\right) \end{aligned} \tag{5.2.8}$$

In Eq. (5.2.8), TP_{ij} is the region of i attractive to j, w_j is the wage rate in the j region, ef_i is the i region of the remote degree, D_{ij} for the i region and the distance from the m region, η as the attenuation factor, P_i is i region of the workforce, and A is the normalized parameter. The relationship of the region is the geographical background of regional growth.

According to Eq. (5.2.8), we obtain each i area to j area labor migration.

$$m_{i,j} = P_i \frac{TP_{i,j}}{\sum\limits_i \sum\limits_j TP_{i,j}} \quad i \neq j \tag{5.2.9}$$

For Eq. (5.2.9), m_{ij} for the i region migrates to the j region of the workforce.

Taking into account the definition of traditional economic growth theory, technology is completely dependent on the labor force, so that the migration of labor will bring the proliferation of production technology. The main performance is production technology related to training of the labor force by the level of education, and the development of the higher ability to spread to other less developed regions; this is a form of spatial knowledge spillover. It is assumed that the R & D innovation capability index of capital organization Agent has the ability to change the numbers in the labor force and the comprehensive technical ability of the labor force; then the innovation capability index of the local area and the innovation ability index of the current population is

$$he_{j,t} = \frac{\sum_i m_{ij} he_{i,t-1} + \left(P_{j,t-1} - \sum_j m_{ij} \right) \cdot he_{j,t-1}}{P_{j,t-1} - \sum_j m_{ij} + \sum_i m_{ij}} \tag{5.2.10}$$

In Eq. (5.2.10), $\sum_i m_{ij}$ is the move into the population, $\sum_j m_{ij}$ is the move out of the population, and $he_{j,t}$ is after the first phase of the population flow, the capital of $t+1$, the Agent innovation capability index.

After loss of part of the population flow in the development and innovation ability of the capital organization, the organization can rely on educational institutions to train the workers to be able to fill the research and development innovation capability. In this chapter, we set the probability of recovery for each research and innovation capacity as

$$P_{he} = A \cdot \exp(I_{\text{edu}} + I_{\text{uni}}) \tag{5.2.11}$$

In Eq. (5.2.11), I_{edu} is the average level of education organizational capital workers could receive, I_{uni} is the index, research, and innovation ability of organizational capital, and A is the normalized index. On the basis of the type, the capital organization of Agent phase t R & D innovation capability is

$$he_t = \min(he_{t-1} + \kappa, he_0) \tag{5.2.12}$$

Here, κ can restore the ability of research and innovation capacity.

Capital Flow

In the model, capital tends to flow to the organization with a higher return on capital, and only the same capital can be obtained. At the same time, capital in the market process should consider the target area of market demand. Market demand is the total demand for the product in other parts of the region. The more rich the target area is, the more people wish to consume and the greater the potential of consumption. Therefore, the numbers of the consumer population and the area of the target region are also attractive for capital flows. We continue to use the Wilson model to characterize the appeal of capital flows. The formula is as follows:

$$TK_{i,j} = w_j P_j \pi_j \exp\left(-\varepsilon D_{i,j}\right) \tag{5.2.13}$$

In Eq. 5.2.13, w_j represents the wage rate of the target capital organization, P_j represents the number of people in the target capital organization, w_jP_j is the consumption scale of the target capital organization, π_j represents the profit margin of the target capital organization, and ε represents the spatial attenuation coefficient of capital flow. Based on Eq. (5.2.13), the capital structure of the capital can be calculated as the largest region max($TK_{i,j}$). Assuming that the capital organization i is the most attractive to capital organization k, $TK_{i,k}$, and $TK_{k,i}$, if $TK_{i,k} > TK_{k,i}$, then in the capital organization i, $k_{i,k}$ is capital organization i and capital organization k, and capital will flow to capital organization k. Capital organization K, capital i capital, and capital organization m capital flow into

$$K_k(t) = K_k(t-1) + \omega k_{i,k}(t) \tag{5.2.14}$$

By Eq. (5.2.14), ω is the loss rate in the process of capital transfer. For the main body of the decision to transfer capital, the remaining capital is

$$K_i(t) = K_i(t-1) - k_{i,k}(t) \tag{5.2.15}$$

Here, t is the current $t-1$.

In this model, in Wang Zheng and other classical analyses of the role of space interaction, as in the particle physics analysis, the particle or charge model of interaction is still set, which cannot be traced back to a deeper level of particle autonomy or adaptability level.

5.2.2 *Simulation Results Based on the Agent-Based Model*

5.2.2.1 Technology in the Space of Diffusion and Evolution Simulation

Based on the model, Wang et al. (2014b) made a Gao Xiang simulation of the technology diffusion and capital flow under different geographical structures. To simulate the formation mechanism of different spatial patterns in the process of technological diffusion and capital flow, that paper analyzes the impact of the road transportation network on technological diffusion and the capital flow pattern. In this chapter, we set two scenarios to reveal the impact of traffic structure on the geographical landscape. In addition, the reason to carry out these two simulations is to show the ABS method, which may handle a complex spatial background. Obviously, the simulation results of the regional road transportation network with the different weights cannot be completed by the analytical method and numerical simulation. Simulation results based on the agent-based diffusion of technology are given in Figs. 5.2 and 5.3.

We can see from Fig. 5.2, in the initial stage of simulation, that the production technology difference between regions is obvious; the central city of higher economic status in the municipality is directly under the central government and the provincial capital city has a higher production technology level, although in other cities production technology level is relatively low. In the simulation, to the 50th stage, the high level of production technology flows from the central city to the surrounding city, forming a center–hinterland structure of the space diffusion circle, the more obvious examples being Beijing and Tianjin and the Yangtze River Delta. From step 1 to the 100th period, with the expansion of the scope of the Yangtze River Delta, and the proliferation of the Yangtze River Delta, Beijing, Shandong, Shanghai, and Jiangsu to a large number of technical radiations from these two core cities, the rapid rise in production technology, the formation of a high-tech belt connecting Beijing and Tianjin, and Beijing and Shanghai, especially the Yangtze River Delta region, has become an integration of production technology. From the 100th to 200th iteration, technological diffusion shows a trend from the east to the middle, the west, and the southeast, with a pattern of double center diffusion. From the whole simulation process, we can see that technology is in the center of a double center: the hinterland space structure, from the center of the city along space linear distance, spreads to the surrounding cities. Because it is a dual-center structure, of

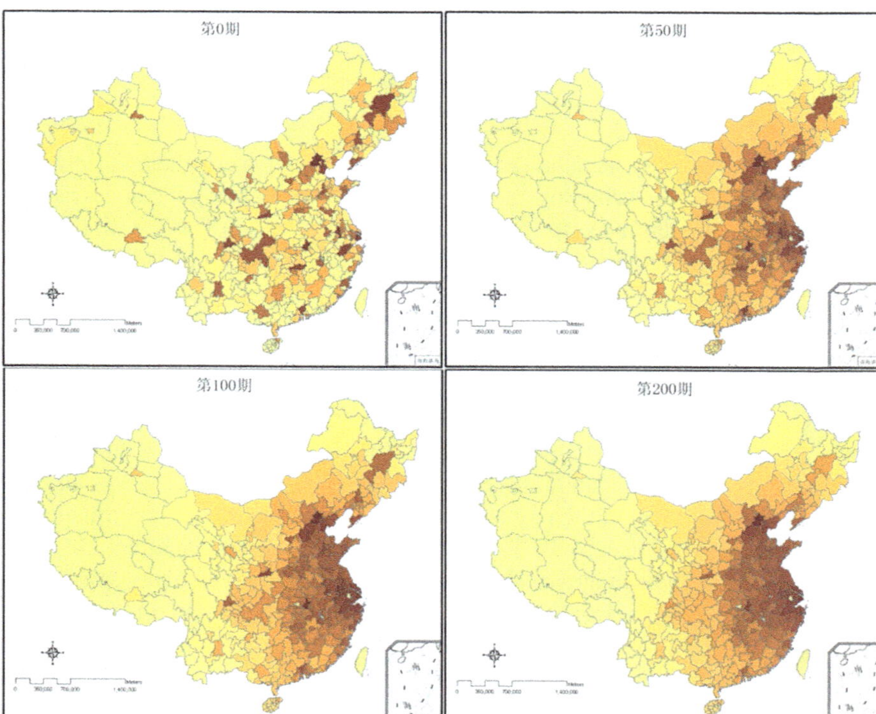

Fig. 5.2 Scenario 1 technology diffusion

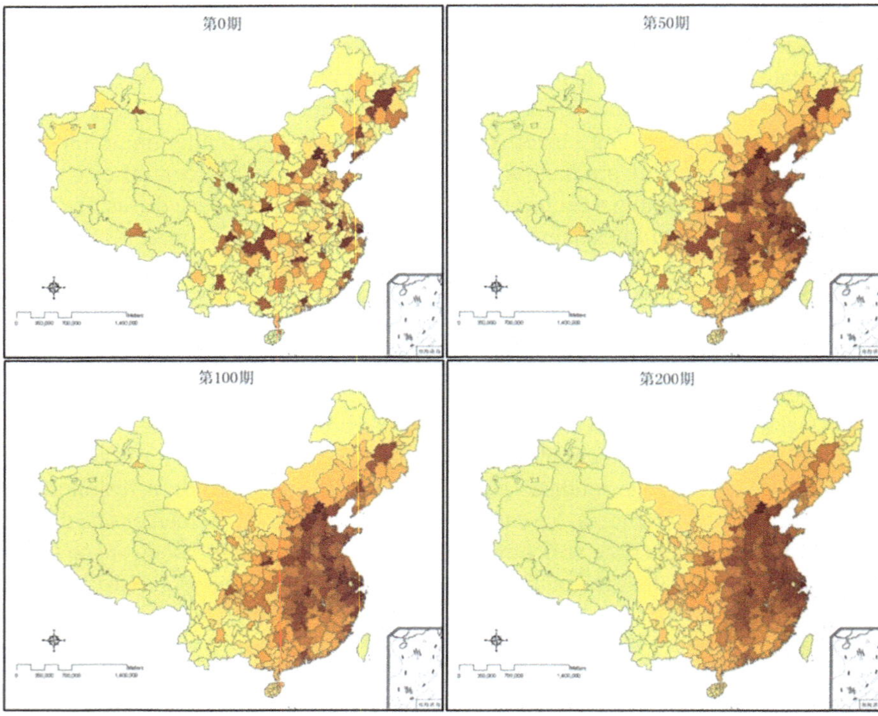

Fig. 5.3 Scenario 2 technology diffusion

the two central cities in the region between the first radiation and the first formation of a higher production technology industry belt, technology along the industrial belt and central cities continues to expand, and in the process of annexation of the smaller central city, gradually forms a cover of the central area of radiation.

From the simulation results of scenario 2 (Fig. 5.3) we can see that the spread of technology is no longer simply from the central city to the surrounding proliferation, but has shown a pattern of spreading along the transportation network; that is, the technology moves along the traffic line so that there is a way out from the center of the city. From the 50th period, this feature began to appear, with high weight, connecting the center of the city with a higher production technology. In the 100th phase, the network main road for the city in production technology is very clear: the higher technology level of the city constitutes an "A"-type distribution (Shanghai–Beijing, Wuhan–Guangzhou–Beijing, Lianyungang–Zhengzhou–Xi'an). In addition, the technical advantages of the Beijing–Qingdao, Beijing–Shenyang–Dalian railway for the city are evident. From the 100th to 200th iteration, the technology has begun to spread from the network backbone to the peripheral branches. In the 200th phase, the technology gap between the central city and transportation network has gradually become smaller, and the gap between urban spaces for Shanghai, Nanjing, Hangzhou, Wuhan, and other cities is not obvious. This development shows that, in scenario 2, the production technology goes from the central city through the network main trunk road, to the main road, and then from the main road node through the network to ordinary nodes.

5.2.2.2 Capital Flow Spatial Pattern Evolution Simulation

Further, from Gu and Wang (2014) and other capital in the simulation of space flows, this result and more can reveal spatial performance. Simulation results are given in Fig. 5.4.

From Fig. 5.4 we can see that the 20th phase results in Beijing, Shanghai, and Guangzhou as three capital hubs to the surrounding city of the capital. Along with the simulation, to the 50th stage, the capital city gradually increased, and showed a trend of uniform distribution. Several important capital cities such as Xi'an, Zhengzhou, Wuhan, Changsha, and other regions have become the regional capital hub, forming the center of its own capital, but most of the capital circles are independent of each other, and no coverage exists. It is worth noting that, to the 50th period, the Yangtze River Delta region has been formed in Shanghai, Nanjing,

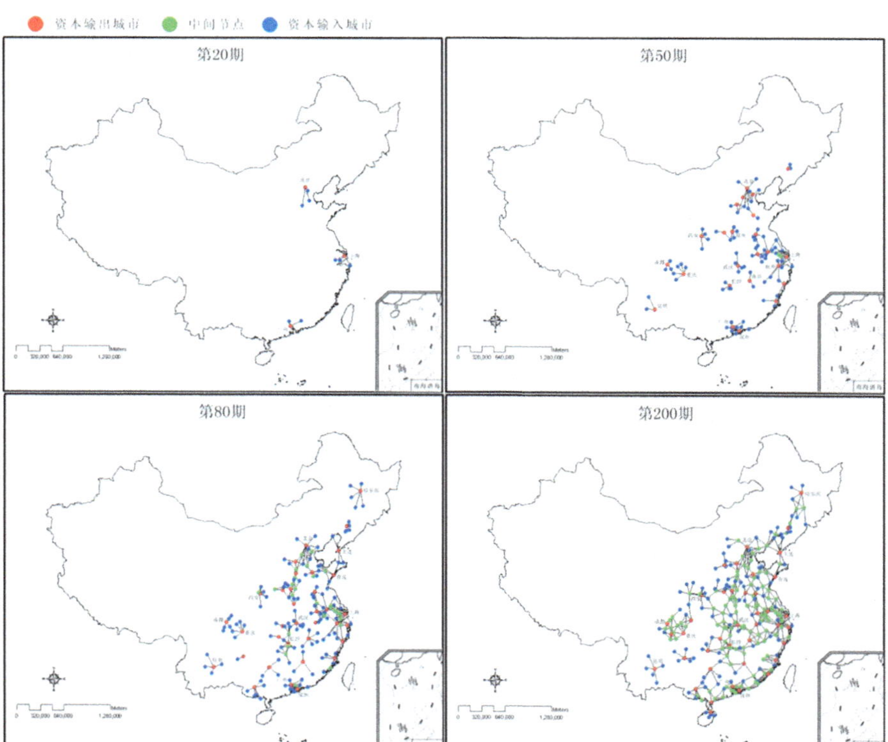

Fig. 5.4 Scenario 1 capital flow distribution

and Hangzhou as the capital hub, with connectivity within each city, a pattern of surrounding radiation, and the emergence of capital flows in the middle node. This result shows that the space network structure is complex. To the 80th period, in addition to the southwest, the northeast regional radiation in the process of expansion is gradually increasing; intermediate nodes have gradually increased, forming a cover of the eastern and central capital flow network. In addition to the Yangtze River Delta region, the edge of the radiation circle is mainly composed of capital input, although the intermediate nodes are located in the middle of the regional capital output hub and the edge of the radiation circle, which shows that the radiation circle is still in a mutually independent state and has formed a central hinterland capital flow structure with Thünen ring characteristics. By the 200th stage, the eastern and central radiation circles are integrated, forming a dense network space structure. At the edge of the original capital input of the city, most of the capital flows into the middle of the node; a part of the original regional capital hub is the capital of the surrounding areas of radiation flowing into the middle node. This scenario indicates that the spatial structure of a multi-center–hinterland capital flow has formed.

Scenario 1 is the capital flow network by a plurality of regional Du Nenghuan independent expansion fusion. In the simulation process, because of lack of interregional capital exchange, it is difficult for the central region to obtain capital radiation from the eastern developed areas.

Based on scenario 2, simulation results (Fig. 5.5) show, from a nonhomogeneous geographical base with a weighted network of roads, that the spatial structure formed by the capital flow is no longer obtained from the scene simulation center–hinterland structure, but a first regional capital hub has connecting roads, with distribution along the main roads of the capital flow network formation, and again by arterial road radiation surrounding the ordinary node hub network spatial structure. In Fig. 5.4, because of the existence of the road traffic network, to simulate the 50th period the regional capital is no longer the standard of the capital hub for the center of the Thünen ring shape, but in the form of channel distribution on the main road (Guangzhou–Beijing, Shanghai–Beijing); these capital flow channels constitute a capital flow network covering the eastern part of the network. In the 80th phase, the capital flow channels connect with each other and with a large number of intermediate nodes, which indicates that the capital flow channels have become the capital flow network formed along the road. At the same time, the network structure is gradually becoming perfect; the capital flow channel along the Longhai line, the Zhejiang-Jiangxi line, began to appear. In the 200th issue, the nodes located on the main road in the capital flow, except for the corolla-embedded capital hub, basically transform as the middle node; the Beijing–Guangzhou, Beijing–Shanghai, Beijing, Kazakhstan, the Longhai, and the Zhe-Gan line-based road network of capital flows has been formed. At the same time, the network structure of the capital flows along the main road to the periphery of the network; the network structure becomes more dense and complex, a process of the development and evolution of network space structure in the capital flow of the road transportation network.

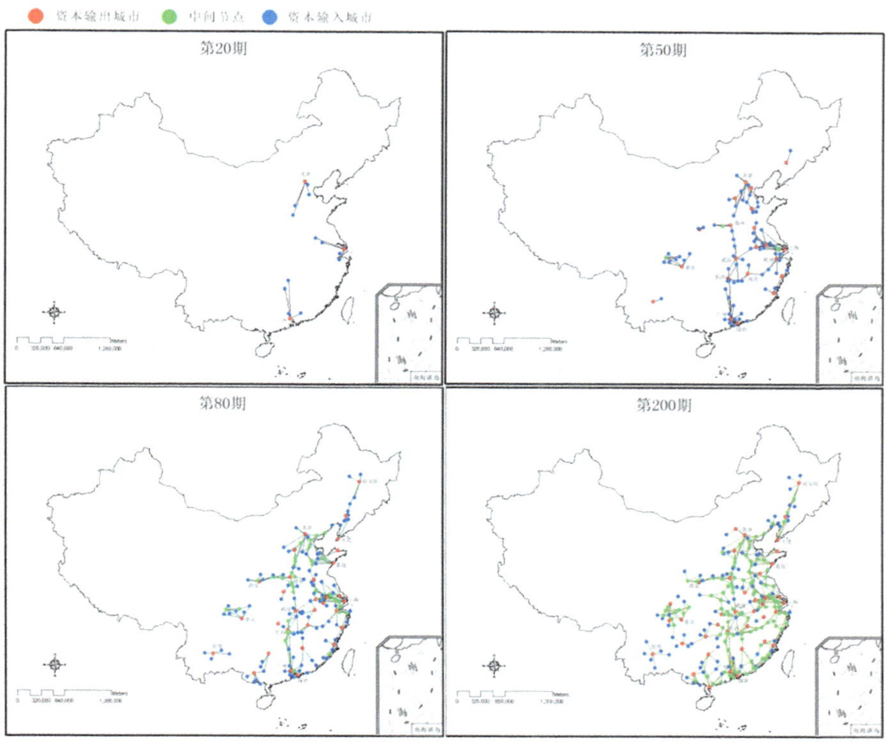

Fig. 5.5 Scenario 2 technology diffusion

Compared to situations 1 and 2 in the simulation of the initial stage of the capital flow, the central region can accept the earlier capital from the East, to strengthen the regional capital flows.

Regardless of situation 1 or scenario 2, simulation of the city space in growth or recession shows that in the city development scale is gradually generated between some nodes from the hub of the network structure, a technical and capital performance contact.

5.2.2.3 Regional Growth Policy Analysis

One advantage of the simulation is that it is easy to carry out a policy simulation analysis. The influence of public policy analysis is a problem of geography, which is an obvious application of color science.

From Figs. 5.2 and 5.3 can be seen the proliferation of core cities, although the greatest technology is close to Beijing, but also a high weight of the network main road (Harbin–Beijing, Dalian–Beijing); production technology in the northeast region has always been very low. Regardless of the use of a spherical distance or a weighted traffic network, the northeast region of technology has always been lower than that close to Hebei, Beijing, Shandong, and other cities. In the introduction of a weighted traffic network simulation, Dalian–Beijing and Harbin–Beijing have not formed a more obvious path with high technology, even to the later simulation; its technical level is greater than Beijing–Shanghai, or Beijing–Wuhan, along which the gap is growing. Because it is generally difficult to set a higher index in the remote northeast region, the labor force with a high technical level is more willing to migrate to the south; to some extent, this hinders technology diffusion by the northeast. To solve this problem, Wang Zheng and other tax incentives (scenario 3) set business income tax and personal income tax in Northeast China to implement 50% of the concessions to attract the flow of labor. By comparing the situation before scenario 2, we obtain the policy analysis conclusion.

Figure 5.6 shows technology diffusion in the northeast without the preferential policies. Simulation analysis, in the case of national unity, shows that production technology level in the northeast has always needed to improve; the ability to accept the core city of radiation is very weak, and the gap between the technological level of Hebei in the simulation process continues to increase. From the 100th to 200th period, Northeast China is able only to better accept technology diffusion to obtain a higher level of technology. After the 200th phase of the simulation, the northeast

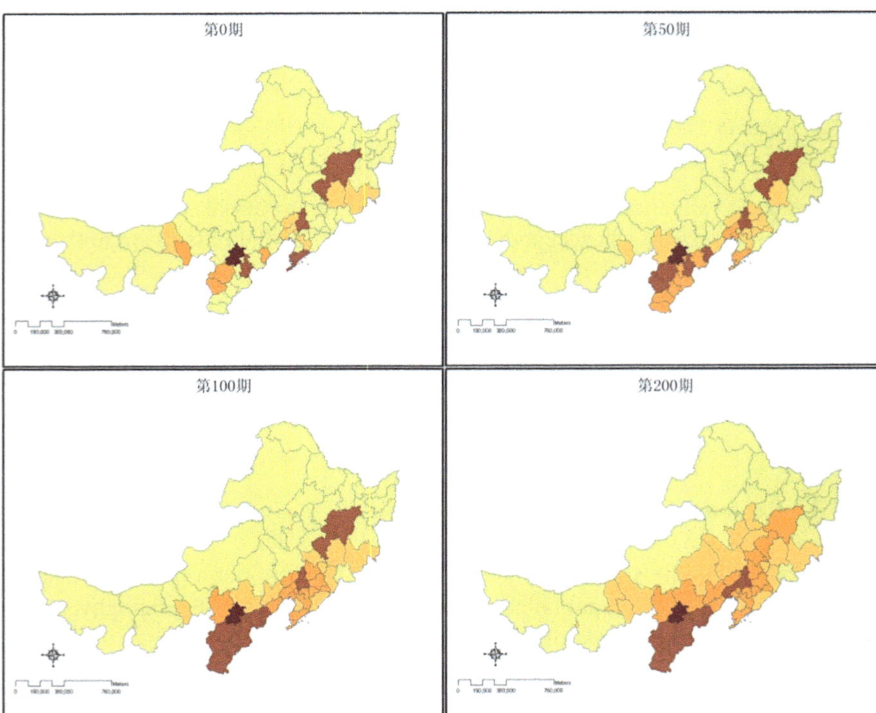

Fig. 5.6 Scenario 2: northeast and other areas of technology diffusion

region has formed technology from the core city of Shenyang to accept radiation by transit and then extending to other areas.

After taking the preferential policies, from Fig. 5.7 we can see that the ability of the northeast region ability to accept the spread of technology has been significantly improved. In the 200th phase, a high-technology industrial belt along the main roads of Beijing–Dalian–Shenyang has formed at the same time as the proliferation from the center of the city to the surrounding urban areas. The technical level along the main road and Hebei, Tianjin, and other approaches shows that preferential policies have a relatively large effect on the migration of labor, which makes high technology a part of the labor force as it begins to migrate to the northeast and enhances the local production technology level.

5.2.2.4 Supplement

Models (5.2.1), (5.2.2), (5.2.3), (5.2.4), (5.2.5), (5.2.6), (5.2.7), (5.2.8), (5.2.9), (5.2.10), (5.2.11), (5.2.12), (5.2.13), (5.2.14), and (5.2.15) describe the basic regional processes, which characterize the major capital and population movements in space, although not yet involved in the innovation activities of enterprises. In fact, regional innovation achieves its profit target life and growth in the behavior of nature. Enterprises inevitably are closely related to the innovation strategy (original and copy/purchase); to achieve innovation, Dai et al. (2008) and Wang et al.

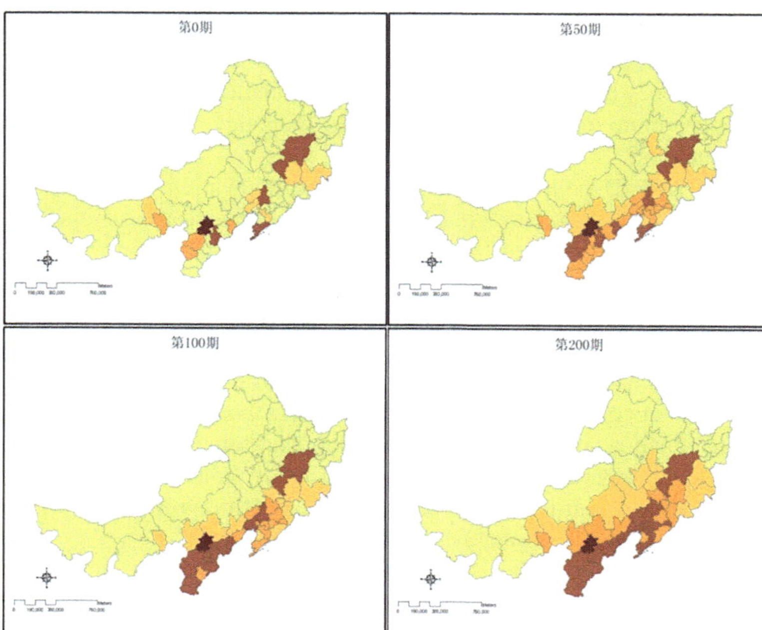

Fig. 5.7 Scenario 3: northeast and other areas of technology diffusion

(2010d, e) developed a model based on agent simulation of multi-innovation strategy behavior. They introduce the formation of technological innovation and emphasize that the spatial distance is the factor that affects the diffusion of technology. The key to the innovation of the model is that there is a technology for the products in the region, and the quality of the product could be improved by λ fold:

$$h_g = h_{g-1} + \lambda h_{g-1} = (1+\lambda)h_{g-1} = (1+\lambda)^{g-1} \quad (5.2.16)$$

where g is the technology generation (grade) in the technical sequence, and h_g is the quality level of the g technology. In addition, in the production level of each generation, its cost decreased to

$$c = \tilde{c}_g \tau^{v-1} \quad (5.2.17)$$

In this formula, c is the production cost of the product, τ is defined as the cost reduction rate, that is, the technical proficiency level, the production cost is reduced, v is the production technology level, and $\tilde{c}_g$ is the standard cost of the g technology. Thus, in the cost of the enterprise, improving the profit is achieved by technological innovation. In this product, innovation behavior is driven by their simulation:

- Regional enterprise group of the initial scale and not enterprise technology development. However, with the continuous growth of enterprises, the capital of agglomeration has effects.
- Regional enterprise group technology entrepreneurship and its development policy tendency have a significant role: initial technical level to high technology progresses more quickly; and the R & D policy tendency is shown as non-monotonic.
- Regional R & D policy orientation (i.e., the innovation of R & D ratio by adjusting the tax) and the regional enterprise technological progress rate in the presence of the downward opening innovation investment proportion show a parabolic relationship between growth and the optimal R & D investment proportion to achieve technology for maximum growth. Moreover, with the increasing protection of intellectual strength, the most gifted new investment proportion is increased, R & D policy being offset to the original.
- Regional intellectual property protection hinders the diffusion of knowledge in the market and is not conducive to improving the technology level of the whole market; it increases the imitation learning difficulty and causes the enterprise to move to independent innovation and technological progress. Enterprise has an optimal R & D policy for innovation input.

5.2.3 *The Formation of Regional Structure in the Space Network*

Work by Wang Zheng et al. shows that the network system will appear in the region. Hojman and Szeidl (2008) established an information game model; their work shows that in the context of the network, the regional structure will form two kinds of stable: center–hinterland structure and star network structure, that is, the hub–network structure. The Hojman and Szeidl (2008) model is as follows:

Consider a simultaneous action of the game; the N node is from the main players, the players to maintain their relationship with each other. At the same time, the collection of the agent-based participants is defined as N. The strategy of autonomous body $i \in N$ is a vector $s^i \in \{0, 1\}^{N-1}$, where $s_j^i = 1$ means that the participants are willing to contact ij; 0 is no contact. All the collection of the self–subject relationship is defined as an undirected graph or network, represented by $g = (s^1, \ldots, s^N)$, where the nodes represent the main body, and the side represents the link.

Each link represents the direct and indirect contact with other agent-based benefits, which depends on the shortest distance between the main body of the$d$ (i,j). Then, the number of nodes connected with $l^i = \sum_{j \in N} s_j^i$ is i, and n_k^i indicates that the number of i is k. The payment function of i is

$$\pi^i\left(s^i, s^{-i}\right) = f\left(a_1 n_1^i + a_2 n_2^i + \cdots + a_d n_d^i\right) - cl^i \tag{5.2.18}$$

Among these, $f(\bullet)$ is an increasing function, and $c > 0$ is the cost of building and maintaining contact. d is associated with the threshold, and if the distance is greater than d, the participants will not receive the proceeds. Hojman and Szeidl (2008) assume that with $a_1 \geq a_2 \geq \ldots \geq a_d$, the distance between neighbors, the greater that the benefits are, the less the existence of distance decay. Further assume that, at this point, direct contact and indirect contact are important. $a_{s+1} = \beta a_s$, and $0 < \beta < 1$, at this time, with the return of indirect contact distance geometric attenuation. The total number of participants we will contact with$\bar{n}^i = a_1 n_1^i + a_2 n_2^i + \cdots + a_d n_d^i$ is i. Further, Hojman and Szeidl (2008) make two key assumptions about the structure of the payment function:

Hypothesis 1: the yield function $f(\bullet)$ is strictly increasing, a concave function, and there is $M \geq 0$, which makes the $m > M$:

$$f(2m) - f(m) < c \tag{5.2.19}$$

The function that satisfies this property shows that (M, c) is strictly decreasing. When $f(\bullet)$ is a bounded function, the c has a M that makes $f(\bullet)$ have a strictly decreasing return characteristic of (M, c).

Hypothesis 2: the communicative threshold d is limited.

Under these assumptions, Hojman and Szeidl (2008) proved that the network system will form a balanced network structure. To that end, Hojman and Szeidl

(2008) defined that if at least one part of the network is maintained, then the equilibrium is not empty. If and for all the links, the subject is known as the center of the collection of the self. If there is an agent-based point that is all self-centered, and there is no connection between the main body and the center, the network structure is the center. In the network, the main body is not connected, the other is maintained as direct or indirect, and $d(j, k) \leq d$, for all $j, k \in N$. The network extent is star shaped. (Obviously, this node is our understanding of the hub.) Thus, Hojman and Szeidl (2008) obtained a mathematical theorem. By this, we can obtain the following proposition.

Proposition 1: if $a_2 > a_3$ and $N > \max(N_0, 4)$, then the only non-null equilibrium is a central hinterland. If $a_2 = a_3$ and $N > \max(N_0, 2d + 1)$, then any one of the non-null Nash equilibrium is an extended star like that of the hub of the network. Among them, $N_0 = (2M)^{2d+2}$.

The proof of this theorem is demonstrated in the strict mathematical formulation of Hojman and Szeidl (2008). Its significance lies in the space base on the network; so long as the existence of the degree of spatial contact with the distance increases in attenuation, it will form a central hinterland structure. If this link does not decay, or if the area is equal, the space will form the hub network structure.

5.3 Spatial Organization Evolution

Geography is different from other natural sciences. It is the result of natural factors and human factors. From the perspective of management, geography is a kind of organization. The spatial organization is the organization form of the landscape image in the geographical space. "Organization" is a kind of ordered entity based on some basic elements. In the organization of human society in the fields of social science, economics, management, and history, their common features are orderly, social, and economic behavior; the spatial organization is the spatial distribution of the spatial distribution of the regional units based on the formation of the ordered solid.

Understanding the geography of spatial organization can be said to be ancient. Before the nineteenth century, Voltaire and Rousseau, in France, formed the enlightenment and rationalism philosophy tradition in Paris as the center of the level of regional morphology: this is a kind of spatial organization. In 1826, they proposed the agricultural location theory. In this theory, the spatial organization of regional development is considered to choose a city as the center of the rural area. In this way, the urban development of rural areas or the use of rural development is the initial regional development management. In addition, under the influence of the French school, in the West the road is the development of the so-called urbanism, this trend of thought emphasizing urban space as the management of the center. In China, "weigh this curb end" has long been stressed, highlighting the rural hinterland of the city, although at the end of the Yuan and Ming twice the city just rose as the center of the path of development for the conservative social thought or a backward nation entered the interrupt. Because of the fundamental influence of the spatial organization in long-term development, it is not stressed that people have the characteristics of the reconstruction, but they are a kind of natural structure. So, sometimes, it has been called the space structure. As a monograph of theoretical geography, this book refers to the spatial structure of space, which is used to refer to the formation of a spatial landscape and the more general spatial organization of human behavior.

In geography, the spatial organization of the region has been summarized as the central place in theory, and the original intention of the central place theory is to explain the formation of the urban system and its hierarchy structure. This theory concludes with the location of the district to develop a center–hinterland structure of development of space and to introduce the model of urban and rural area binary structure matching regional development management. The core of the center is the integrated function to provide products and services to its hinterland area, whereas the hinterland is the basic element of the support center to provide the raw materials, labor, and other basic elements. The space center of the organization model is the center of political rights or economic geography, and the periphery is the hinterland. In the later development, the hinterland is known to be divided into two levels, the up-conversion belt and the down-conversion belt. In the presence of a large number of centers in space, it forms a hierarchical structure, which is a senior center and a low-level center. This theory has brought about the establishment of the regional production complex in regional development management, and the idea of dividing the economic zone, each center and hinterland matching the regional production complex, in the competition between the central city and the city, and the competition has brought about the regional innovation power. There is no doubt that this kind of spatial structure theory can promote regional post-war development.

This center–hinterland is the organizational form of the current popular geopolitical economics theory. As is known, Machinder in 1904 published the geopolitical theory, emphasizing the "center of the Eurasian continent." This is the edge region of the heart, a kind of center–hinterland structure. Earlier, in the second century BC, the historian Sima Qian put forward a similar view of China's geopolitical economic structure: "so the land of Guanzhong, in the world, and the people of the three; however, the amount of its rich, even its six.... Cardiff Sanhe in the world, if the king, the more in also." It is worth noting that in Sima Qian, Mackinder is the center–hinterland of organizing space, "if the three legs of a tripod, King more habitat also" conclusions may be drawn from history. In other words, it is a kind of organization theory.

Under the influence of location dynamics, the formation of the Chris Taylor mode is still the center of the structure of the hinterland, but it has evolved into a more complex internal structure, so that the so-called market principles of the three central shared or equally divided structures form the hinterland, as well as a large center controlling six small centers and then controlling the hinterland of the organization structure (Fig. 5.8).

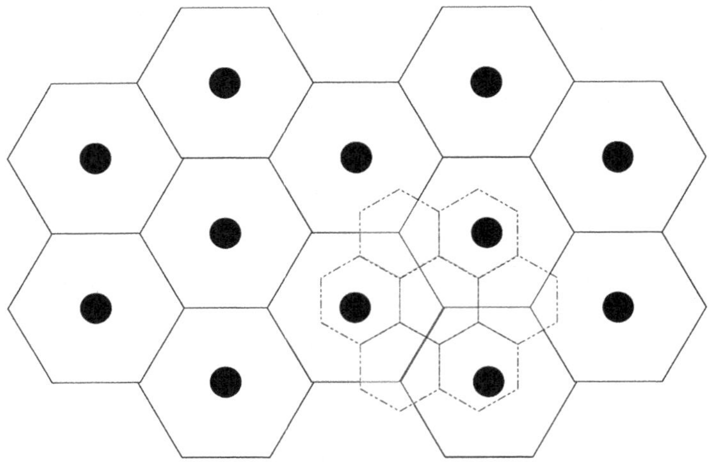

Fig. 5.8 Center of Chris Taylor theory structure and its corresponding center organization form

In China, Lu (2001) proposed urban residents: all types of transportation and energy supply lines, information lines, etc., in the space (region) should be applied to organize the needs. His opinion is that the space unit could be organized as a point–axis model. Under this kind of organization structure, some cities have become the growth center; they have their own professional status, the transportation line, the logistics channel to "axis" in the form of connection, strengthening economic ties, which have become the growth axis, and no longer just the van Thünen position to pull the economy. This pattern of organization structure, the control of the center of the hinterland, demolished the foundation of the "regional production complex." In fact, under the condition of the market economy, product logistics demands that the traffic line is free to seek the market. The local logistics and the enterprise can choose their own specialized direction if they are not willing to be a central city in the economy. This phenomenon occurred in Zhejiang Province; to adapt to economic development, revocation occurred in the "region" of this level of administrative units. In Henan Province, the industrial cluster, there are many local industrial clusters or "villages"; they are directly serving foreign companies and bypass the central city of the market. In this case, there are a number of cities to become professional under the market, business, research, and development hub, local nodes and their links to the formation of network: the location of the structure cannot exist among them. To develop the industry, the R&D institute of the enterprise can be far away from the center city. For example, Shanghai is a research and development hub, with its research and development services, and information services, such as the Yangtze River Delta region and Shaanxi; Yunnan is such a remote region: connecting the hub and the local node is not the vast hinterland and is a kind of traffic network, an information network. That is to say, the new economic form and information, and other factors, promote the space to organize the network in the form of a network.

In general, the hub is also a center, but it is not inclusive of all external control. It is generally believed that the hub has a professional function and the center has more emphasis on comprehensive functions, that the hub is the network rather than the hinterland, and the network has a number of nodes affecting the surrounding areas. In addition, the hub is connected with others, and it is likely to be cooperative, which is a fundamental difference from the center–hinterland structure. Between each center in the center–hinterland structure is competition and striving for maximum coverage. Once the structure has achieved stability, between the same levels of formation equilibrium, the secondary center in the superior center at the boundary finds the remaining space, until the new conditions appear to change the balance of power. However, there is a lack of effective communication and exchange between the center and the different level centers, resulting in the isolation of the center, the formation of a closed center–hinterland unit. Because of the lack of cohesion between regions, each also can achieve self-sufficiency at a low level; therefore, the center–hinterland structure needs the powerful central authority support, because once the central power has weakened, very often this may lead to regional separatism. In China, the Xin Jiang autonomous provinces before and after the revolution of the separatist state are typical. In contrast, the hub network structure connects fully; although this connection may be directly connected, or may be indirectly connected, but it must be fully connected, that is, at least there is a path between any arbitrary hubs, the hub of connectivity for the professional division of labor. After the division of labor, the service objective of the hub is not its own peripheral or hinterland in urban and rural nodes, but rather to cover all other hubs and nodes in the whole region, and also the need to receive other hub products and services. Frequent hub transactions and a huge regional traffic flow will require the hub is effectively connected. This kind of connectivity is more expressive in the hub position. A city may have a pivotal position in the direction of a certain industry, its economic spillover affecting the whole country, but in another direction, it may only be a common node, receiving radiation; this avoids the struggle at the entrance of local competition. Third, the bearing capacity of the city has regional leadership and not only an administrative center position or production center position; the main function of the hub is a regional spillover. Such spillovers include knowledge spillovers, market spillovers, and resource spillovers. Knowledge spillover here refers to the hub of the city as a research and development center, driven by technological change and network coverage of the welfare growth; a cultural innovation center brings cultural change and the resulting network benefits. The market spillover is the hub of commerce and trade functions, and technology, especially the proliferation of technology standards to benefit from the proliferation of resources; resource development is the key to the development of other network nodes that may bring the market location to technological interests; in short, regional leadership and economic spillovers. This spillover may be a small place to create a hub, such as Yiwu, Zhejiang, Baigou, and Hebei having become hubs in a professional sense, with leading markets of small commodities of regional leadership. In fact, in the Pearl River Delta of Guangzhou and Shenzhen, Guangzhou is a leader in the regional industrial hub city; Shenzhen is a leading national electronic industrial city. In one area, they have organized the Yangtze River Delta.

O'Kelly et al. (1996) proposed two possible hub network structures, and their research shows that when the hub transport conditions deteriorate, the node cities may be connected with other cities and replace the hub, and even select two or more hubs. The node with only the hub of the network is known as the single distribution network, whereas the nodes and multiple hubs-connected network is called a multi-distribution network. The existence of this multi-distribution hub makes the innovation of the whole space structure increase with the growth and development of the region. Under the new economic conditions, the general development of a multi-distribution network is seen in Fig. 5.9.

The hub network structure was initially recognized in the area of the port system by van de Kilink (1998). The port is a special kind of place; it has the functions of transportation, trade, and distribution, and around the port a special logistics system will form. This logistics system makes the local nodes of the region have different functions, forming different levels of different shapes of nodes. Wang et al. (2002, 2012b) concluded that there are four kinds of spatial units in this structure.

(1) Hub and common node: in the storage space, population, and capital in a certain place to gather, the aggregation of the population point, an economic place has a certain geographical feature; this is the node. These nodes are the basic units of the network structure. In a port system, the port and goods storage, living areas, and docks are the nodes and the port center is a logistics hub, as a flow of material and a gathering place, usually in the early development of the ports. Ordinary ports are also common nodes. In the general economic system, capital-intensive hub functions are regional R & D activities, trade activities, and/or production activities, densely populated areas, or a gathering center. In the region, the local and regional nodes are not in the hub, but it is relatively small; it provides labor support and other economic support to the hub and network and has lower production function, financial capacity, and some special function with general professional influence. In the administrative system, these are political centers at all levels: this is a scenic

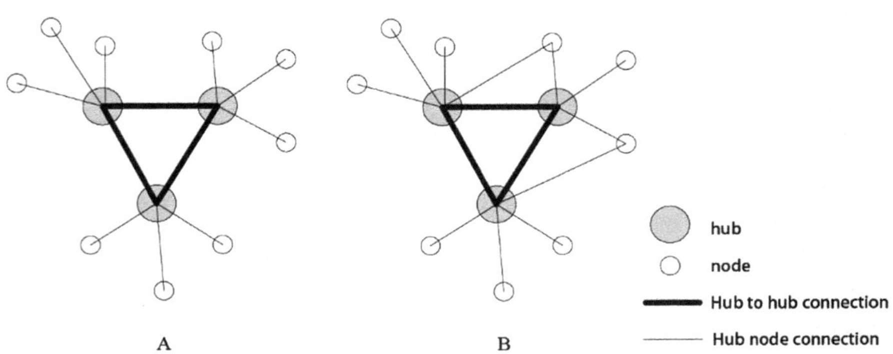

Fig. 5.9 Two-hub network structure (O'Kelly and Bryan 1998): single-distribution network (**a**) and multi-distribution network (**b**)

spot in the tourism area. In all, it emphasizes that the entire system of nodes, to the hierarchical hub sequence, fills the entire area, such as the capital, capital city, and the county-township residents. This node system forms the evolution of regional hubs in different regions.

In particular, it is important to emphasize that the hub is a specialized division of labor; a port may be a logistics hub, but it is not a regional manufacturing hub and cultural hub. Beijing, for example, is clearly a cultural and political hub, but in the international trading system, it is a logistics or trade hub, and Tianjin is more in this regard. Of course, in railway logistics, the Beijing hub status is much higher than in Tianjin, which means that the hub is different from the center, but with a functional division of labor. In China's electronic industry system, Shenzhen is clearly more important than Nanjing; Hangzhou has a higher position, because much of the trade and innovation occurs here, and in the traditional manufacturing industry, Nanjing, Hangzhou, and Shenzhen are more than a hub; the same is true of Hangzhou, in the national leisure industry, and a prominent position. The hub center is not a master; it is a product of specialization.

(2) The core and the attachment region: the hub is a specific space, the core area of the implementation of the basic functions of the hub, the auxiliary function of the core area; the problem is attachment to the hub of the surrounding areas. Of course, we can also see the two functional areas, the "domain of" title. van de Kilink (1998) observed, in the port system, that the port functions with the development of industry and its complex. The space scale of the port has expanded to the adjacent areas, in addition to the distribution along the pier, but also includes the industrial area, which is not a port as with the constraints. At this stage, the hub functions as a space to differentiate the functional units, and these units are attached. There is almost no independent economic behavior; the hubs provide support for the core functions of the entire project. In a port as a hub, people observe that the increase in equipment and capital becomes intensive; the core area is the center of the structure and contains the region of several domains. In the general industrial economic system, the capital-intensive industry nodes are the center of the formation of a certain kind of logistics flow in relationships to people. In the supply of land and of labor, the nodes provide the basis for facilities and functions in the system of a tourist area, such as the formation of a scenic area with a certain traffic, an accommodation area, and an essential position in the community. What really has regional significance is the scenic area, but the transportation and accommodations participate in the construction of the hub of the included area. In China's electronics industry, we can observe an electronic industrial innovation hub in the core periphery where, in the area beyond Baoan of the Shenzhen area, the manufacture of samples completed the pilot industry of the region. Further Dongguan, in fact, also implemented a similar function of Baoan. Thus, Shenzhen is the hub of the electronics industry, which needs innovation, research, and development as its main features. Shenzhen has become China's R & D hub, in addition to the business functions of Futian, but Nanshan is the research area of Shenzhen, for the development of a field but also for improvements.

(3) Hub cycle: the hub of the area or the node as the center of the surface area is the top of the hub and a sub-hub adjacent to the formation of a central area. This core area is a central place in a certain position; it can realize the function of a certain area, and other sub-areas and the regional characteristics of the key areas are integrated. Shanghai as the core hub, with Hangzhou, Suzhou, Wuxi, and Changzhou/Ningbo as secondary hubs in the formation of China's manufacturing hub, but it also highlights the status of Shanghai's manufacturing innovation hub. It is worth noting that Hangzhou, Suzhou, Wuxi, Changzhou, and Ningbo are not a repeat of the functions of Shanghai but rather its professional aspect. It is worth noting that this hub may cover the hinge of the administrative territories, it may be beyond the original administrative area, but also includes the hub of the territory around the place, as does the hub of Shanghai; its hub, including Zhejiang and Jiangsu, also includes a number of nodes that have a certain level of specialization. In this process, the area of the node of the hub is associated with knowledge spillovers and GDP spillovers.

It is worth noting that, in this circle, the development of a number of marginal areas as a hub of the professional park. Where the edge of the hub is a specialized capital-gathering place, its working population generally does not live in this same place; hence the need for public services, and trade services, as a product of the modern economy. China's typical metropolitan area is the hub of Beijing; with Tianjin, Tangshan, and Langfang as the center of the hub.

(4) The distribution system: the connections of the hub network structure and the general node need a circulation system organized as a system of nodes. The most direct distribution system of supporting regional logistics activities of the transport system, which is a fast way under local economic conditions, is the road system relative to having connections for the region. In the economy developed along a certain zone will form a dense industrial distribution with a connecting hub or secondary hub, such as the high-technology industry development after the Beijing to Tanggu, London to Reading, and Guangzhou to Shenzhen industrial zones. Of course, in the modern economy, the connection between the nodes may rely on modern fiberoptic cable, radio/dependence, and non-material contact, so that the hub of the network structure is not a fourth kind of space form.

However, this is a network connection in an abstract space; some people call it cyberspace. And in China, it is translated into the network space, which is very easy to confuse with the network. The 2006 edition of the standard vocabulary of geography translated it as "collateral space," emphasizing that it is a kind of road space in mathematics.

From these four factors, the network is a hub network structure. Such as a number of ports and the mainland can form a port network, then the port area is the function of the hub. It is worth noting that different functions may exist in the same hub. Also, in other economies, different functions of the hub are found.

In the hub and network structure, the hub has the function of leadership. Under normal circumstances, the hub of the region's major business or organization activities is the industry development direction, the impact of the entire region. Within the hub of innovation, an innovation first in the hub of the node stimulates the proliferation of nodes, rather than the space of the neighborhood: capital and population aggregation, but also the main point adjacent to the upper level, rather than the central place. Similarly, movement at this level also follows the flow of logistics, the hierarchical movement of the hub network structure, which makes a distinct difference between the center and the hinterland.

As we have seen, in the center of this discussion, the structure of the center–hinterland can be nested, and the network structure could also be nested. As a place, the CBD (central business district) is the hub, the business district is the node, and the main road is the formation of a hub–network structure of the road system. It must be noted, however, that because of the limited nature of the actual geographical space the center–hinterland structure and hub–network structure cannot be infinitely nested.

In fact, in wastelands, the hub network is always in its own regional structure. In a wasteland, because of the existence of certain settlements of human activities, the settlers may establish settlements in the core areas of the original development to obtain resources, and thus they become a space population and an economic node. Where there is a core area of the road link, then there will be a star-like structure development. At the same time, a new node between the horizontal link, the network structure, will thus be formed. In contrast to the general spatial network, the settled population is bound to carry out economic activities and trade, so the development of the population-economic nodes will not only strengthen their own node function. In further innovation and trade, some nodes will be active in the city, a regional structure in the traffic that develops, in trade volume, a cultural and innovative and emerging hub of the city. Since the Middle Ages in China, the development of the Yangtze River Delta region has undergone such a historical process. It is noteworthy that in the Yangtze River Delta, the initial hub city is Suzhou. However, since the nineteenth century, as a result of the development of international trade, the accumulation of regional capital, and the impact of colonization, Shanghai has replaced Suzhou as a regional hub; in other words, external forces may change the structure of the region.

Chapter 6
Multi-regional Economic Processes

Abstract One of the characteristics of this book is the discussion of multi-regional processes, because the world is composed of multiple regions. The main three multi-regional economic models are shown here. The first is the Mundell-Fleming system; the second is the Wang and Gu (2015) global economic integration model, improved by Jin (Am Econ Rev 102(5):2111–2146, 2012); and the third model is the global geopolitical game model, which, of course, appears simple. The behavior of multi-regional economies and the geopolitical economic game are directions of regional science that need further research.

Global economic integration is the trend of the world economy in the late twentieth century. Global economic integration is regional economic development, the customs union; the common market has gradually developed. Wang et al. (2002) discussed the theory of economic geography, this development process. Because of space limitations, this book only discusses global economic integration, the operation of the global economic system.

Global economic integration has two models: one is based on the spillover of gross domestic product (GDP). The earliest model is the Mundell-Fleming model, which is a classical model. The other is based on the global economy in general equilibrium, taking into account the practical link between the global economies. The model developed by Jin (2012) is such a model. The model and its theory are still developing.

6.1 Mundell–Fleming System

6.1.1 GDP Spillover

The common model of multi-GDP spillover is based on the Douven model described by Mundell-Fleming and Douven and Peeters (1998). The M-F model usually contains two regions: national and foreign; Douven and Peeters (1998) extended the model from two regions to multi-regions. This chapter includes the international interaction model of China, the United States, Japan, Russia, the European Union, India, and other countries, in which the countries interact with

Z. Wang, *Principles of Regional Science*, New Frontiers in Regional Science: Asian Perspectives 15, DOI 10.1007/978-981-10-5367-2_6

each other, and GDP spillover influences the mechanism. GDP spillover refers to the economic interaction between countries through trade, technology, capital output, markets, and other mutual influences. In theory, the model has several key assumptions, namely, capital flows, money market equilibrium, national income identity, exchange rate fluctuations, and interest rate market. But in practice, people have to do with historical parameters.

The Mundell-Fleming model of the system is as follows: each region has a total demand curve, the total supply equation (or Phillips curve), and a money demand equation. The consumer price level is defined as the weighted average of domestic and foreign prices. Because of the number of countries and the number of equations, the form of the equation is the same, but the country is different.

Equation (6.1.1) is a national C standard LM curve:

$$m_C - p_C{}^c = \omega_0 + \omega_1 q_C - \omega_2 i_C + \omega_3 \left(m_{C-1} - p_{C-1}^c\right) \tag{6.1.1}$$

Here, -1 indicates that the lag value is $p_c = p_c(t)\,, p_{c\,,\,-1} = p_c(t-1)$; the variables in the equation, in addition to the short-term loan interest rate i, are all natural logarithms. The parameters are assumed to be positive. m is the nominal monetary equilibrium; p^c is the consumer price index; q is the real GDP; i is the short-term loan interest rate; and $\omega 1$, $\omega 2$, and $\omega 3$ are the coefficients of the equation.

Equation (6.1.2) is the actual aggregate demand equation that is interpreted by the actual interest rate, nominal interest rate, foreign GDP, government expenditure, and time trend. In fact, it represents the general demand of a country and the real exchange rate.

$$q_C = v_0 + \sum_{j \neq C} \left(v_j \lambda_j + v_j' q_j\right) - v_c\left(i_C - p_{C+1} + p_C\right) + v_C' G_C + g_c t \tag{6.1.2}$$

λ is currencies against the U.S. dollar real exchange rate; $+1$ represents a numerical; p_j is price level; G_C is the national government spending; ν_j, ν_j' is for j on its influence coefficient; g_c is for no GDP spillovers; and C is the intrinsic rate of increase. The second relationship characterizes the interaction between countries through spillovers.

Equation (6.1.3) describes the consumer price of the weighted average of the commodity price and the price of imported goods.

$$p_C^c = \rho_1 p_C + \sum_{j \neq C} \rho_j\left(p_j - e_j + e_C\right) \tag{6.1.3}$$

e_j is the exchange rate of various currencies against the dollar, and $\rho_1\,, \rho_2\,, \ldots\ldots \rho_n\,, n \in C$ is the coefficient of the equation.

$C =$ real exchange rate:

$$\lambda = e_C + p_A - p_C \tag{6.1.4}$$

Here, p_A is the price level in the United States, because we use the U.S. dollar as the standard currency.

The model building is based on the theory that the parameters will be based on the statistical estimation of the measurement method, and the model can reflect the real economic openness of the countries in the world. In the world, other than the considered economic body can be used as a residual item, so long as the statistical test can be considered as a model. Parameters of the model might be statistical estimates by using the regression method.

Based on the equation system (6.1.1), (6.1.2), (6.1.3), and (6.1.4), Huang et al. (2015) use the data of China, the United States, Japan, Russia, the European Union, and India 1990–2009 to estimate the parameters.

From China's equation, the GDP of US has a positive spillover effect; the coefficient is 0.5396, and the coefficient of Japan is −0.2444, which is the result of the competition between GDP of China and GDP of Japan. It is a significant positive spillover effect of GDP. From the United States equation, the GDP of the United States and the European Union GDP have a substantial negative spillover effect (significance, 0.0506 and 0.0362, respectively); which is the same as the financial industry in leading countries; the GDP of China and Russia have a more obvious positive spillover effect, and the result is more significant, which is the fundamental basis for the cooperation between China and Russia. The EU equation shows that the GDP of the United States and Japan have a negative spillover effect on the EU GDP (significance = 0.135 and 0.1557), China's GDP has a positive spillover effect on the EU GDP, and the results are very significant (significance = 0.0767). This is the cause for the dominant presence of friction in the financial sectors of developed countries. From Japan's equation, the GDP of the United States and the European Union have a negative spillover effect on Japan's GDP, the result being −1.0724 (significance = 0.3219) and −4.8468 (significance = 0.2619); which is more obvious; the spillover effect of China to Japan is 1.2703 (significance = 0.2747). China's economic development is favorable to the Japanese economy, which was been recognized by the Yukio administration in those days.

6.2 Global Economic Integration Model

The multinational GDP spillover model fails to consider the balance of supply and demand in the international economy: the world may not be in a general equilibrium state. Jin (2012) developed a model trying to avoid this problem.

In the Jin (2012) model, the whole world is been divided into several countries, in the following, to show that each country has a production department. The model assumes that the capital and goods are free to flow, there is no trade barrier, there is no flow of labor among the countries, and a free flow of labor within the country. The behavior of consumers by overlapping generations are the model to describe the enterprise; the capital stock is determined by the period of investment.

A. Production Behavior

The Jin model defines the production behavior of the product in the way of the Cobb-Douglas production function. Jin (2012) assumes that the same department has the unique output elasticity; the same country has the sole production technology level; and the production technology level is completely effective in the labor force.

$$Y_{i,t}^{j} = \left(K_{i,t}^{j}\right)^{\alpha_i}\left(A_t^{j} L_{i,t}^{j}\right)^{1-\alpha_i} \tag{6.2.1}$$

In Eq. (6.2.1), $Y_{i,t}^{j}$ is the country's i sector j in the first phase of output, $K_{i,t}^{j}$ is its fixed capital, $L_{i,t}^{j}$ is the number of laborers, A_t^{j} is the production technology level, and α_i is the output elasticity of the department. In this paper, by Arrow and Debreu (1954) and Romer (1986), with the assumption of the production technology level and the department of capital accumulation, the expression is as follows:

$$A_t^{j} = B_t^{j}\left(K_t^{j}\right)^{\varphi} \tag{6.2.2}$$

The labor wage rate is equal to the marginal product of labor, which is the same as the wage rate of different departments in the same country.

$$w_{i,t}^{j} = \frac{\partial Y_{i,t}^{j}}{\partial L_{i,t}^{j}} p_{i,t} = (1-\alpha_i)\frac{Y_{i,t}^{j}}{L_{i,t}^{j}} p_{i,t} \tag{6.2.3}$$

Here, $p_{i,t}$ is the price of product i, resulting from the absence of trade barriers between countries, so the market price of a product within a period is unified. According to Abel, in Jin (2012), capital is updated in Cobb-Douglas form.

$$K_{i,t+1}^{j} = a\left(I_{i,t}^{j}\right)^{\phi}\left(K_{i,t}^{j}\right)^{1-\phi} \tag{6.2.4}$$

In Eq. (6.2.4), $I_{i,t}^{j}$ is the number of investments in the t sector of the state j i, including domestic investment and foreign investment in two parts, ϕ is the output elasticity of capital production, and $a$is the total factor productivity. $I_{i,t}^{j}$ (6.2.3) for the partial derivative of the investment, the marginal output of the investment, is the reciprocal of the price of capital:

$$q_{i,t}^{j} = 1/\frac{\partial K_{i,t+1}^{j}}{\partial I_{i,t}^{j}} = \frac{1}{a\phi}\left(\frac{I_{i,t}^{j}}{K_{i,t}^{j}}\right)^{1-\phi} \tag{6.2.5}$$

Enterprises in each GDP consists of two parts, namely, production and capital production:

$$\mathrm{GDP}_t^j = \sum_i p_{i,t} Y_{i,t}^j + \frac{1}{\phi} I_t^j - I_t^j = Y_t^j - \frac{1-\phi}{\phi} I_t^j \tag{6.2.6}$$

B. General Equilibrium Conditions and Market Clearing Conditions with Equilibrium

Product market clearing conditions are equal to the sum of the output and the consumption and investment, namely with the current consumption of the total value quantity and the next period capital stock total value and equal. In the period of the assumption global output is $Y_{i,t}^g$, then the product clearing condition for i:

$$Y_{i,t}^g = \sum_j Y_{i,t}^j = \sum_j c_{i,t}^j + \frac{1}{\phi} \sum_j \sum_i x_{k,i,t}^j \tag{6.2.7}$$

According to Jin (2012), the relative relationship between product price and output:

$$\frac{p_{i,t}}{p_{k,t}} = \left(\frac{Y_{k,t}^g}{Y_{i,t}^g} \cdot \frac{\gamma_i}{\gamma_k} \right)^{\frac{1}{\theta}} \tag{6.2.8}$$

In Eq. (6.2.8), γ_i represents the share index of each product. To simplify the equilibrium model, this paper uses two basic hypotheses of Jin (2012): (1) it is assumed that the consumer goods, the investment, and the price of the composite functions are often replaced by $\theta = 1$; and (2) it is assumed that consumers have a logarithmic utility preference, that is, $\rho = 1$. In this way, the equation can be simplified as clearing prices:

$$p_{i,t} = \frac{\gamma_i}{Y_{i,t}^g} Y_t^g \tag{6.2.9}$$

In addition, the labor clearing equation can be expressed as:

$$\sum_i L_{i,t}^j = L_t^j \tag{6.2.10}$$

C. Investment Pattern Design

By the general equilibrium condition, the wage equation and the price equation can be obtained:

$$W_t^g = \sum_j \sum_i (1-\alpha_i) Y_{i,t}^j \frac{Y_t^g}{Y_{i,t}^g} \gamma_i = Y_t^g \sum_i (1-\alpha_i) \gamma_i \tag{6.2.11}$$

A fixed share of total wages for all visible market output to $\sum_{i}(1-\alpha_i)\gamma_i$, accounting for the proportion of total output is the number of each investment market:

$$I_t^g = \phi \sum_j s_j \sum_i (1-\alpha_i)\gamma_i Y_t^j \tag{6.2.12}$$

In Eq. (6.2.12), s_j is the savings rate of j.

In the real world, the international capital flows are very complicated because of the influence of the saving behavior and economic policies of the various countries, and this phenomenon is often difficult to express in the two models. Therefore, Gu and Wang (2014) developed a multinational model of Jin (2012), assuming that the flow of capital is composed of two parts, which has both the pursuit of the size of the economy and the pursuit of capital return, so that the proportion of investment in the total investment of all sectors of the country:

$$R_{i,t}^j = \mu R_1(i,j,t) + (1-\mu) R_2(i,j,t) \tag{6.2.13}$$

In Eq. (6.2.13), R_1 and R_2 are in the pursuit of economies of scale and the proportion of investment in the pursuit of capital return ratio; $\mu \in [0, 1]$ is the weight coefficient of R_1. The expression of R is derived from Jin (2012), and the basic idea of the design is to ensure that the return of the unit investment in various sectors of the world is equal to that of the various sectors of the world:

$$R_1(i,j,t) = \frac{\alpha_i \gamma_i}{s_k + \frac{(1-\phi)\beta}{1+\beta} s_l} \mathrm{E}\left[\frac{Y_{i,t+1}^j}{Y_{i,t+1}^g}\right] + \frac{\frac{(1-\phi)\beta}{1+\beta} s_l}{s_k + \frac{(1-\phi)\beta}{1+\beta} s_l} \mathrm{E}\left[R_{i,t+1}^j\right] \tag{6.2.14}$$

In the form of Eq. (6.2.14), the R_1 is a capital allocation model that takes the output of the investment. In the calculation of the model, we use simulation results obtained by $\bar{Y}_{i,t}^j / \bar{Y}_{i,t}^g$ and the average $\bar{R}_{i,t+1}^j$ instead of on the next expected E $\left[Y_{i,t+1}^j / Y_{i,t+1}^g\right]$ and $\mathrm{E}\left[R_{i,t+1}^j\right]$.

R_2 is based on the attractiveness of the capital investment accounted for by various sectors of the country. Different from R_2, R_1 from the microscopic point of view takes a bottom-up view of the capital flow in the process of profit, that is, the high rate of return on capital sector flows. As the global economic integration is the emergence of regional capital flows, the use of regional (national) capital stock, the total number of wages, and the marginal output of capital are factors of capital attraction strength. Wang and Ge (2002) used the exponential form of the area of the output difference between regions, the regional capital attraction strength formula:

$$TK_{i,j}^{x,y} = K_{i,t}^{j} w_t^{y} L_t^{y} \frac{\alpha_x Y_{x,t}^{y}}{K_{x,t}^{y}} \exp\left(-\upsilon \left| \ln \frac{Y_t^{j}}{Y_t^{y}} \right| + 1\right) \tag{6.2.15}$$

$$R_2(i,j,t) = \sum\nolimits_{x,y} \frac{\frac{TK_{x,y}^{i,j}}{\sum_{i,j} TK_{x,y}^{i,j}} \left(\frac{\phi \bar{\beta}}{1+\bar{\beta}} S_{x,t}^{y} \right)}{I_t^{g}} \tag{6.2.16}$$

In Eq. (6.2.16), $TK_{i,j}^{x,y}$ is the country's y sector x to the national j sector i capital attraction strength, υ is the impact of the value of capital increase ratio of the two countries, $S_{x,t}^{y}$ and the savings by the national y sector x.

The various obstacles and policies in the international capital investment mean that capital investment is not likely to be complete. In this paper, we use the investment weight to characterize the impact of policies on capital free flow, to overcome the error of simulation. In this way, the investment ratio could be expressed as

$$R_{i,j,t}^{*} = A\xi^{j} R_{i,t}^{j} \tag{6.2.17}$$

In Eq. (6.2.17), A represents the normalized parameter. The number of investments by various countries in the world is

$$I_{i,t}^{j} = R_{i,j,t}^{*} I_t^{g} \tag{6.2.18}$$

and its current account is expressed as

$$CA_t^{j} = w_t^{j} L_t^{j} - \sum\nolimits_i I_{i,t}^{j} \tag{6.2.19}$$

Application of the Jin (2012) model results in space, in Wang et al. (2014a, b).

6.3 Global Geopolitical Relations

6.3.1 Overview

The geopolitical relationship is the geographical distribution of the government, the economy, and the ecological community because of the interests of the relationship and the formation of the fate. In this kind of predestination, the earliest attention is given to the relationship of political and military aspects, especially the control of the geographical unit. The relationship eventually led to these geographical units to form the strategic alliances, quasi-covenant, and hostile relations or policy, according to Chinese "take the words too literally," understanding; this kind of predestination is the "geo", the geographic fate. In the twenty-first century, with the development of world economic integration, the economic ties between countries are becoming more and more close, and the mutual influence is growing. Since the U.S. financial crisis of 2008, the whole world is subject to shocks. The impact of the crisis on the different degrees of the world, is constantly changing the world, but also changes in the international political and economic pattern. In fact, the U.S. Defense Department Senior Consultant, in Washington, the strategic and international relations research center of strategic expert Root Walker (N. Luttwak Edward) in 1990 spoke at a congressional hearing for the first time. Edward believes that the world has entered the era of geopolitical, economic, social, ecological, and other issues, the urgency and importance of the first time significantly more than the "senior politics" (military confrontation and nuclear deterrence), whereas the geopolitical strategy of the military confrontation has a buffer role, this in turn to open up a new era of more than national boundaries, competition, and cooperation. In addition, Demko and Wood (1994) is a further combination of geopolitical and economic competition and put forward the geopolitical economics (geopolinomics, or geopolitical economics).

In early geo-politics, Alfred T. Mahan put "sea-power theory," Halford and Mackinder put the "land power theory" and "heartland theory," which is famous, greatly affected. Sevilla Kanchelskis (de Seversky) put "air power theory," Nicholas J. Spykman put "edge theory," and Cohen (1991) added the new geo-politics. In modern political and economic conditions, these theories need to be inherited and developed.

It must be pointed out that the geopolitical issues have traditionally been used to be misunderstood: first, it must be global; second, it must be political. Actually otherwise from geo-politics, the earliest in China formed a "vertical," a "horizontal"; they are not about the world. Of course in China at that time, the emperor was weak without political authority, or "vertical"; "horizontal" is a kind of international relationship theory and policy. However, after the unification of China, Sima Qian put forward another theory: "the land of Guanzhong, in the world, and the people of the three; however, the amount of its rich, even the six, ... In the Tang Dynasty are East, Yin people of Hanoi, people of Henan. Cardiff Sanhe in the world, if the king, the more in also." Sima Qian's theory, emphasizing "in the world, if the three legs of a tripod, King more home", and Mackinder's world island–heartland theory are similar. Also, Sima Qian's theory is considered to be a country's geopolitical relations or geopolitical economics.

Machinder, in the completion of his discussion, put forward the idea that he respected the geopolitical issues, that is, four research principles:

- The impact of geography on the country can be sought from history;
- The outline of the world is clear;
- The world has three kinds of different concepts of land right and sea power;
- Scientific and technological progress has provided the opportunity to achieve a large modern state at that time.

Here Machinder emphasized the historical method, and Sima Qian is a great historian; it is easy to find that, similar to the geopolitical case, the only thing that happens in the world is not likely to be the case in the case of empirical testing, and a need to use historical experience. Of course, Mackinder's point of view, particularly the last one, has an obvious political tendency. Modern geography has stepped into the "pure science" era; the conclusion of the study is not political. This is also the emphasis of this book: that the geopolitical economics is different from the traditional place of geo-politics.

6.3.2 New Worlds Model

Under the framework of geopolitical structure theory, Wang proposed that the world can be divided into three different types: financial owners, manufacturing owners, and the owners of the supply of resources, referred to as the financial, manufacturing, and resource industries, respectively; in fact, these three countries formed three different geopolitical domains. Based on this idea, Xiong and Wang (2012) by building the corresponding model system tried to simulate and explore the new world geopolitical economics.

In the Xiong and Wang (2012) model, the world is no longer divided into the hinterland structure, but the world economy contains three kinds of mutual dependence and mutual control of the state: A, B and C, which represent the financial, manufacturing and resource countries. Three factors of production: capital, labor and resources; and three sectors: manufacturing, agriculture, and services. The flow of labor is within the department, not in the international flow. We assume that the total population of A (B, C) is $L_A(L_B,L_C)$, and the effective labor population is $\varsigma_A L_A(\varsigma_B L_B,\varsigma_C L_C)$, $\varsigma_A(\varsigma_B,\varsigma_C)$ is A (B, C). The national capital endowment is held by the local national average, and the per capita capital of its residents is $k_A(k_B,k_C)$. Capital in the international flow of capital: A (B, C) of capital holders can be invested in A, B, C three countries of the enterprise. A (B, C) can be used to $N_{SA}(N_{SB},N_{SC})$, the resources for each country is homogeneous, the resources can flow in the international.

A. Resident

According to Pflüger (2004), each country residents in i preference is[1]:

$$U_i = \alpha_i \ln C_{M,i} + \beta_i \ln C_{S,i} + C_{A,i} \tag{6.3.1}$$

Among them, $\alpha_i > 0$, $\beta_i > 0$, $C_{M,i}$ said the i country's manufacturing product consumption set, $C_{M,i}$ said the country's service industry product consumption,

[1]Pflüger (2004) proposed the semi logarithmic preference of residents, the model considers the senior high school entrance examination two departments of agriculture and manufacturing industry. In this model, the model is extended to the three sectors of agriculture, manufacturing and service sectors, and that the service industry is similar to the manufacturing industry, not the basic consumption, and the existence of diminishing utility.

$C_{A,i}$ said the i country's agricultural product consumption set, $i = A, B, C$. The nature of the country is different; the parameters are not the same.

For $C_{M,i}$ we assume it is the following form:

$$C_{M,i} = \left(\sum_j x_{ji}^{\frac{\sigma-1}{\sigma}} \right)^{\frac{\sigma}{\sigma-1}} \tag{6.3.2}$$

Among them, $\sigma > 1$ said the substitution elasticity between different countries in the manufacturing products, x_{ji} said the j country's consumer spending on the country's manufacturing products, $j = A, B, C$.

Residents consumption budget:

$$G_{M,i} C_{M,i} + p_{S,i} C_{S,i} + p_{A,i} C_{A,i} = C_i \tag{6.3.3}$$

Here, $G_{M,i}$ is the price index of CES in the form of i, $p_{A,i}$ is the price of agricultural products, $p_{S,i}$ is the price of service products, C_i is the i country for the consumption of the part of the disposable income per capita.

$$C_i = (1 - s_i) y_i \tag{6.3.4}$$

Here, s_i is national savings rate. y_i is per capita disposable income of Chinese Residents of i.

According to Pflüger (2004), the model of price index$G_{M,i}$ design for the following form:

$$G_{M,i} = \left(\sum_j \left((m_i/m_j) \tau_{ij} T_{ij} p_j \right)^{1-\sigma} \right)^{\frac{1}{1-\sigma}} \tag{6.3.5}$$

p_j is state production of manufactured goods FOB. $m_i(m_j)$ is exchange rate of i and j. Units the country i (the country j) Currency per National world currency exchange$m_i(m_j)$.τ_{ij} is iceberg cost coefficient from the country i to the country j, which is loss of transportation cost. Per unit of i country (J) product output to the j country (i) only $1/\tau_{ij}$ units of the product. When $i = j, \tau_{ij} = 1$.T_{ij} is i tariffs on products of j. $T_{ij} - 1$ is i tariff rate on the products of j. Therefore, $(m_i/m_j)\tau_{ij}T_{ij}p_j$ is j country enterprise production of manufactured goods in the country market sales price. We assume that the technological level of agricultural production is influenced by the production technology level of the manufacturing industry, and it depends on the input of labor, and the agricultural production function[2] of i is

[2]We believe that the improvement of the level of manufacturing technology can improve the agricultural labor production level, however, the specific functional relationship between them is not clear. The setting of the agricultural production function is just a simplified treatment of the

$$X_{A,i} = A_i L_{A,i} \tag{6.3.6}$$

$X_{A,i}$ is agricultural output of the country i. $L_{A,i}$ is agricultural labor input of the country i. A_i is manufacturing technology level of the country i.

We also assume that the agricultural products as the basic consumer goods, all kinds of agricultural products are homogeneous, in the country's self-sufficiency, so in the case of complete competition, agricultural products price is equal to its marginal cost, labor price is equal to the marginal output of the product. The agricultural products as a benchmark commodity, thus determining the i country's wage rate$w_i = A_i$. At this time, the total demand of the country i for agricultural products is $L_i C_{A,i}$.

In addition, for service products, service is entirely the labor of workers, we think a unit of service corresponding to a unit of labor, so the service industry production function of the country i set to

$$X_{S,i} = L_{S,i} \tag{6.3.7}$$

$X_{S,i}$ is service output of the country i. $L_{S,i}$ is service labor input of the country i. At this time, the service products price of i is

$$p_{S,i} = A_i \tag{6.3.8}$$

Referring to Yamamoto (2008), the utility of the i country's consumer demand function is:

$$C_{M,i} = \alpha_i G_{M,i}^{-1} \tag{6.3.9}$$

$$C_{S,i} = \beta_i p_{S,i}^{-1} \tag{6.3.10}$$

$$C_{A,i} = C_i - \alpha_i - \beta_i \tag{6.3.11}$$

$$x_{ji} = \alpha_i \left((m_i/m_j) \tau_{ij} p_i \right)^{-\sigma} G_{M,i}^{\sigma-1} \tag{6.3.12}$$

In Eq. (6.3.8) substitution Eq. (6.3.10) demand for service products in the country I residents, and we can get I country residents of service product per capita demand, and then shows the total demand of the service products is $L_i \beta_i A_i^{-1}$.

In addition, according to the demand of the national consumer to the manufacturing industry, we can also get the total demand of the i country's manufacturing products:

core problem of the model. Below the setting of the service industry production function, also encountered similar problems, we also made the corresponding simplified treatment.

$$X_{D,i} = \sum_j \tau_{ij} L_j x_{ij} \tag{6.3.13}$$

$\tau_{ij}L_jx_{ij}$ is the total demand for the manufacturing of *j* country by *i*. Due to the loss of transportation, the *j* country residents consume 1 units of *i* products, *i* countries to provide τ_{ij} units of the product.

B. Enterprise

According to Krugman (1991a, b), the production of manufacturing products, including fixed costs and constant marginal cost, and the existence of scale economy:

$$F_i = \gamma_i + v_i X_{MS,i} \tag{6.3.14}$$

F_i is *i* production of goods in the use of intermediate products input. $X_{MS,i}$ is output of manufacturing products of the country *i*. γ_i is fixed cost. v_i is marginal cost.

Production function of intermediate product:

$$F_i = A_i \left(\xi_{L,i} L_{MD,i}^{(\rho-1)/\rho} + \xi_{K,i} K_{D,i}^{(\rho-1)/\rho} + \xi_{N,i} N_{D,i}^{(\rho-1)/\rho} \right)^{\rho/(\rho-1)} \tag{6.3.15}$$

$\xi_{L,i}$ is distribution coefficient of labor force, $\xi_{K,i}$ is distribution coefficient of capital, $\xi_{N,i}$ is resource allocation coefficient, $L_{MD,i}$ is actual labor force in manufacturing industry of the country *I*, $K_{D,i}$ is actual investment in manufacturing industry of the country *i*, $N_{D,i}$ is actual resources of manufacturing industry of the country *i*, and ρ is substitution elasticity of elements.

So the manufacturing function of country *i* can be written as

$$\begin{aligned} X_{MS,i} = {} & -\gamma_i / v_i \\ & + v_i^{-1} A_i \left(\xi_{L,i} L_{MD,i}^{(\rho-1)/\rho} + \xi_{K,i} K_{D,i}^{(\rho-1)/\rho} + \xi_{N,i} N_{D,i}^{(\rho-1)/\rho} \right)^{\rho/(\rho-1)} \end{aligned} \tag{6.3.16}$$

It is assumed that all the enterprises in the world can be produced according to the optimal allocation of resources and the cost is minimized. $C_{T,i}$ is total cost of production of intermediate products of the country *i*.

$$\min C_{T,i} = w_i L_{MD,i} + r_{K,i} K_{D,i} + p_{N,i} N_{D,i} \tag{6.3.17}$$

$$s.t.\ \ F_i = A_i \left(\xi_{L,i} L_{MD,i}^{(\rho-1)/\rho} + \xi_{K,i} K_{D,i}^{(\rho-1)/\rho} + \xi_{N,i} N_{D,i}^{(\rho-1)/\rho} \right)^{\rho/(\rho-1)} \tag{6.3.18}$$

According to the principle of minimizing cost:

$$C^*_{T,i} = A_i^{-1}\left(\xi^\rho_{L,i} w_i^{1-\rho} + \xi^\rho_{K,i} r_{K,i}^{1-\rho} + \xi^\rho_{N,i} p_{N,i}^{1-\rho}\right)^{1/(1-\rho)} F_i \tag{6.3.19}$$

$C^*_{T,i}$ is total cost under the optimal portfolio, w_i is labor remuneration (wages), $r_{K,i}$ is return on capital (interest rate), and $p_{N,i}$ is price of resources. It is assumed that the production of the intermediate products in China is completely competitive, so the price is equal to the marginal cost, and thus the price index of intermediate products is equal to the marginal production cost.

$$p_{F,i} = A_i^{-1}\left(\xi^\rho_{L,i} w_i^{1-\rho} + \xi^\rho_{K,i} r_{K,i}^{1-\rho} + \xi^\rho_{N,i} p_{N,i}^{1-\rho}\right)^{1/(1-\rho)} \tag{6.3.20}$$

So, the profit function of i country manufacturing enterprises can be written as

$$\pi_i = \sum_j \left(p_i - v_i p_{F,i}\right)\tau_{ij} L_j x_{ij} - \gamma_i p_{F,i} \tag{6.3.21}$$

Manufacturers pursue profit maximization; according to Krugman (1991a, b), then the country i will be the factory price set in

$$p_i = \frac{\sigma}{\sigma - 1} v_i p_{F,i} = \frac{\sigma}{\sigma - 1} v_i A_i^{-1}\left(\xi^\rho_{L,i} w_i^{1-\rho} + \xi^\rho_{K,i} r_{K,i}^{1-\rho} + \xi^\rho_{N,i} p_{N,i}^{1-\rho}\right)^{1/(1-\rho)} \tag{6.3.22}$$

C. Market Equilibrium

In the case of market equilibrium, supply of the country i is equal to the demand $X_{MS,i} = X_{MD,i}$. Because of the optimal allocation of i manufacturing enterprises in the production, the production rate of the situation, according to the principle of minimum cost production, i = the manufacturing industry, the input of various factors of production.

$$L_{M,i} = \left((\gamma_i + \nu_i X_{MD,i})/A_i\right)\left(\xi_{L,i}/w_i\right)^\rho \left(\xi^\rho_{L,i} w_i^{1-\rho} + \xi^\rho_{K,i} r_{K,i}^{1-\rho} + \xi^\rho_{N,i} p_{N,i}^{1-\rho}\right)^{\rho/(1-\rho)} \tag{6.3.23}$$

$$K_{D,i} = L_{MD,i}\left(w_i \xi_{K,i}/r_i \xi_{L,i}\right)^\rho \tag{6.3.24}$$

$$N_{D,i} = L_{MD,i}\left(w_i \xi_{N,i}/p_{N,i} \xi_{L,i}\right)^\rho \tag{6.3.25}$$

The market equilibrium not only needs to ensure that the product market clearing, but also includes capital market and resource market. Because the capital flow in the world is free, so the capital market clearing by Eq. (6.3.26) is

$$\sum_i \left((I_i + k_i)L_i - K_{D,i}\right)/m_i = 0 \tag{6.3.26}$$

I_i is per capita investment of the country i, and $((I_i+k_i)L_i-K_{D,i})/m_i$ is the balance of capital supply and capital demand of the country I in the world currency.

$$I_i = s_i y_i + k_{Credit,i} \tag{6.3.27}$$

$k_{\text{Credit},i}$ is credit capital.[3] For clearing the resource market, by Eq. 6.3.28:

$$\sum_i (N_{S,i} - N_{D,i}) = 0 \tag{6.3.28}$$

$N_{S,i}-N_{D,i}$ is resource supply and demand balance of the country i. For the labor market, according to Keynes's effective demand theory, we allow the unemployed:

$$L_{UnEM,i} = \varsigma_i L_i - L_{MD,i} - L_{A,i} - L_{S,i} \tag{6.3.29}$$

$L_{UnEM,i}$ is the unemployed population of the country i, $L_{A,i}=X_{A,i}/A_i$, $L_{S,i}=A_i X_{S,i}$.

D. Simulation Parameter Settings

Fujita et al. (2001) pointed out that for a three-region model it is difficult to obtain the analytical solution of the three regions, we but can use the numerical simulation method to obtain some numerical solution, which is used to analyze the three regions. Therefore, this model is implemented by C# and Matlab. The core part of the optimization algorithm for the model uses Matlab as prepared and uses the C# to achieve the call interface. Thus, Economy Simulation System Three-Country (T-CESS) is developed as an analytical tool. The development process of this is skipped.

Using the T-CESS system to carry out the economic dynamic simulation, we need to set the initial values of the parameters and variables. The specifics of the model of the exchange rate changes in the simulation, the relevant parameters, and the initial value of the setting are found in Table 6.1.

To characterize the three types of financial, manufacturing, and resource countries, the model has a 4,001,600,800 population. It mainly reflects the abundance of a manufacturing country, and 0.7, 0.6, 0.55, which reflects the level of economic development of the three countries, which reflects the differences in the level of technology, and the 1.4, 1.3, and 1.2, respectively. The model considers that the

[3]There is a huge difference in the level of financial development in the financial development level of the financial state, the manufacturing country and the resources industry. With its highly developed financial market and strong financial strength, the financial state can exert great credit creation ability, strengthen the domestic investment ability, while the manufacturing country and the country of resources industry are lack of the ability to create credit, and the ability to convert into investment is also weak. For this purpose, we introduce the credit capital, which is used to express the financing capacity of the three countries. If, shows that the country's financial markets are developed, the country's financial market is lagging behind, the savings investment conversion rate is low.

Table 6.1 Simulation parameter settings

Parameter	Value	Parameter	Value	Parameter	Value	Parameter	Value	Parameter	Value	Parameter	Value
ρ	1.6	σ	2			k_A	0.7	k_B	0.6	k_C	0.55
y_A	1.5	y_B	1.4	y_C	1.3	s_A	0.2	s_B	0.3	s_C	0.3
L_A	400	L_B	1600	L_C	800	C_{AA}	0.2	C_{AB}	0.2	C_{AC}	0.2
ς_A	0.5	ς_B	0. 5	ς_C	0.5	N_{SA}	150	N_{SB}	600	N_{SC}	1500
α_A	3.2	α_B	3.1	α_C	3	m_A	1.2	m_B	5.1	m_C	5.6
β_A	0.55	β_B	0.5	β_C	0.4	A_A	1.4	A_B	1.3	A_C	1.2
γ_A	1000	γ_B	1000	γ_C	1000	ξ_{LA}	0.49	ξ_{LB}	0.52	ξ_{LC}	0.56
ν_A	1	ν_B	1	ν_C	1	ξ_{KA}	0.16	ξ_{KB}	0.2	ξ_{KC}	0.24
τ_{AB}	1.1	τ_{AC}	1.1	τ_{BC}	1.1	ξ_{NA}	0.35	ξ_{NB}	0.28	ξ_{NC}	0.2
T_{AB}	1.2	T_{AC}	1.2	T_{BA}	1.2	T_{BC}	1.2	T_{CA}	1.2	T_{CB}	1.2

Note: The value variables associated with variable initialization are expressed in their own currencies savings rate is lower, and the savings rate of the financial, manufacturing, and resource countries are 0.3, 0.2, and 0.3, and the basic exchange rate is 1.2, 5.1, and 5.5, respectively; The model assumes that well-developed financial markets have a strong credit creation ability, whereas the manufacturing country and resources of credit lack creative ability, the transform rate of saving to investment is low, and the country's financial, manufacturing, and resources of credit assets were set to 0.36, −0.08, and −0.09. The model assumes that in different countries the consumption of agricultural products as the basic consumer goods is the same, at a onetime capital setting of 0.2; the cost of the iceberg among the countries is 1.2, the tariff coefficient is 1.2, the capital allocation coefficient is set according to the capital, the capital adequacy ratio is 0.16, the manufacturing country is 0.2, and the resource country is 0.24. The resource allocation coefficient is set according to the degree of resources; the most scarce financial resources are subject to resource so the maximum is set to 0.35. The production of a certain degree of resource abundance of state-owned resources is slightly less than 0.28; if resource rich, it is less than 0.2. The labor distribution coefficient is 1 minus all other resource allocation coefficients, specifically 0.49, 0.52, and 0.56, which in a sense is a reflection the degree of a skilled labor force, reflecting the countries producing the same product of labor consumption. For other settings, see Table 6.1 and reference related variables in the model description.

D. Simulation Results

Xiong and Wang (2012), through simulation analysis, found the following conclusions:

1. The devaluation of any country will bring about the improvement of its output and employment, but it will lead to a decline in output and employment in other countries. The output and employment of the financial industry is less affected by the exchange rate change; the domestic production and employment in the manufacturing sector are relatively large, whereas the output and employment of the resources industry are sensitive to the changes of exchange rate.
2. The exchange rate is a two-edged sword, for any country, through devaluation, will bring home an improved employment situation but will reduce the level of per capita consumption of their own people; appreciation will reduce the employment rate, but can bring the rise of the level of consumption of their own people. Therefore, in the adjustment of the exchange rate of the country, the essence is the choice of the employment rate and the consumption level.
3. In the political and economic game of the customs, the financial industry has the absolute right to say; only the combination of the manufacturing and the resource industries can make an impact on the formation of the financial industry. Resources are in a completely inferior position in the tariff game, but it can be through the joint response of each country to seek its own economic interests.
4. The constraint of resources has a restriction on economic growth; increasing the intensity of resource constraints will slow the growth of the economy of the country, and the manufacturing industry is more vulnerable to the impact. At the

same time, with the increase of resource constraints, the risk of economic crisis caused by the financial industry in the manufacturing sector is also greater. And the intervention of the price of resources, reducing the price level of the financial industry and manufacturing countries, will further delay the development of the economy, and increase the volatility of the changes of employment, output, and consumption.

5. The credit consumption structure of the financial industry is an important reason for the financial crisis. The collapse of the credit system in China has caused a sharp decline in consumer spending, which has brought the country's financial crisis, and further spread to the world.
6. In the event of a global economic crisis, the political and economic game of exchange rate and tariff is not the fundamental way out of the crisis. Increasing the supply of public goods, and through the transfer of subsidies and other means improving the capacity of consumption by residents, will help us out of the economic crisis.
7. For the manufacturing country, the employment loss caused by the implementation of the currency appreciation after the crisis is less than the normal economic operation period, and the increase of the per capita consumption of the residents is greater than the normal economic operation period. That is to say, if the currency of the manufacturing sector needs to be appreciated, it is more advantageous for the manufacturing country to choose to be more advantageous than the normal economic operation period after the outbreak of the crisis. In the context of economic crisis, the manufacturing country to a certain appreciation, and with some of the government's public goods supply, transfer payments and other domestic regulation and control of employment losses, so as to achieve the improvement of the level of domestic consumption, is a viable option.

6.4 Discussion

The geopolitical theory, whether Sima Qian's geo-economics theory or Mackinder's heartland theory, essentially is a center–hinterland structure theory. In the new three worlds in the model, the author's thinking is no longer a center–hinterland structure, but a relationship of the equality of the hub–network structure. (In the seventh chapter, see Hojman and Szeidl 2008.)

It has been shown that space is the basis of the network; so long as the existence of the extent of spatial contact is attenuated; it will form the center–hinterland structure. If this link does not decay, or if the area is equal, the space will form the hub–network structure. Under the condition of the development of modern science and technology, all countries in the world need economic integration to guide their professional division of labor status; in general, the world overall is equal. In other words, the development of science and technology has brought the development of geoeconomic structure to such a level that the world can only be equal; the new world may be the result of specialization, so that some countries become the hub of a certain type of industry, the center of the formation of a star network. As the industry is diversified, the industrial hub can also be diversified. The theory of Cohen (1991) has already broken through the center–hinterland structure theory. Cohen (1991) stated that the world is actually divided into a geopolitical structure that is a hierarchical structure; the first level is the geographical strategic area (geo-strategic realms). He divided this into two fields: the global ocean domain and the Eurasian land. Below the domain is the geopolitical area. The domain is the competition power unit, the district is the space continuity, and as the political, cultural, military, and economic interaction points out, its formation is strongly influenced by history. In general, the district is a subordinate unit of the domain, but there are three zones that are not included in the domain: South Asia, Middle East, and central and Eastern Europe are considered to be independent of the domain. The third level is the state. Rbelt shatte is also a conflict zone, which he believes in the 1990 years of broken belt in the Middle East, these zones result from the role of the two major regions of the country to become a prominent conflict between the regions. The Channel (Gateway) area is a peaceful transition zone; the channel region is balanced in two domains, which are the two domains of economic transmission, social and cultural transmission; the zone is open to becoming the two domains of the junction. In the foregoing system, Cohen is also divided into quarter-share marginality (of); the world in the marginal zone is independent from the economic system, and cannot benefit from the changes in the interests of the two domains. Cohen put forward the theory of the era, that the world has not completely complied with the center–hinterland structure of the organization. The reason lies in the development of technology to bring the development of developing countries; they require more equal opportunities. At this time, in the center of the geopolitical structure are the signs of the collapse; for those who are independent of the zone, the broken belt of the country's economy is to fight for national equality. In fact, in September to December 1991, the Soviet Union as the core of the geopolitical domain officially collapsed. Cohen's article was published in the first issue of the American Association of Geographers journal (AAG) in 1991. In a changing world, according to the Hojman and Szeidl (2008) theory, the world will inevitably have a hub–network structure.

Wang et al. (2014a, b) discuss the center–hinterland structure and the hub–network structure difference (compared in Table 6.2); this is not to say that it is a better structure, but a statement of evolution characteristics. As the information society brings many social connections, the hub–network structure develops. The achievement of this development to the geopolitical thought is that the so-called center of the world is no longer able to control other countries, because of the full information available; the division of social collaboration is more important than the regional competitiveness. Knowledge spillovers have a role in the new world.

In the current world, the financial industry, the manufacturing industry, and the resources industry have formed: each country has its own national interest. Because of the development of science and technology and human civilization, a country

Table 6.2 Comparison of two regional development management spatial organization models

	Hub–network structure	Center–hinterland structure
Political logic	The system of equal development of flat	Hierarchical system
Economic thought	Specialized division of labor under economic integration	Regional production complex
Economic society	Information economy era	Industrial economic times
Regional relationship	Division of labor and regional leadership	Pay attention to regional competitiveness and control
Service function	Overall specialization and integration	Local comprehensive functions
Factor support	Second nature, third nature	First nature, second nature

trying to "the sage holds yuan," the quartet to the center hinterland structure has been outdated. Here, the factors of human civilization are not to be ignored; Mackinder then sees the role of science and technology, so that the "opportunities in large modern state" control the heart to the possibility of the world organization, has confirmed that the history of the development is wrong; the diversification of the mode of human civilization, the center hinterland structure challenges; the clash of civilizations become in the world the main conflict, and the conflict leads to three geographical natures of national industrial division of labor based on type. In the modern world, it is not possible not to use Arabia's and Russia's energy, and maybe not China and India, the manufacturing products; of course, the world is dependent on the country's financial industry capital and high-technology products, if a civilization mandatorily forced the world to accept their model, the world may have incessant wars. Of course, this is not to say that the world is to tolerate some of the local anti-civilization models, but to promote cultural penetration, because by relying on the financial advantages, manufacturing advantages, and resource advantages brought about by the decay of civilization by distance, the world may be in the center of civilization that is still preserved in the hinterland structure. And in the world in this case, to avoid the conflict among the three types of civilizations, it is possible to avoid war. In particular, we have tried to break the different types of national industrial supply chains, or the world will be on the outbreak of war. In recent years, as the financial industry as the national cultural center of a country actively returning to Asia, inciting some lust for countries trying to cut off China as a world manufacturing industry center in the supply of resources, is forcing China and other manufacturing countries and resource industry countries to unite. Similarly, in the clash of the basis of civilizations, some Catholic countries lay siege to Orthodox countries, which has caused the edge of the country's civil war. The world has placed Ukraine, Poland, Syria, Iraq, and other countries in the Cohen-like broken belt. In the present world, the financial sector is being wrapped in financial sanctions or resources, destroying the original stability of the world, which has always been the world's fire; on the other hand, some traditional forces for the traditional pursuit of interests and geopolitical thinking are trying to rebuild themselves as a geographical center. The hinterland of the center, in challenging and curbing the growth of national industrial centers, is promoting new crushing areas. The country is pushing a new broken area. How to keep peace in the world, while maintaining human civilization, is a common challenge facing humankind, but also the task of geopolitical economics.

Chapter 7
Simulation Methods for Regional Economic Development

Abstract The final chapter of this book, which introduces the computable general equilibrium model (CGE) method, is an appendix of regional analysis and the methods of multi-regional DCGE analysis. Written by the authors from their many years of experience, it can be used by regional economic researchers.

7.1 CGE Analysis of Single Regional Economy

7.1.1 Introduction of the CGE Model

The computable general equilibrium (CGE) model describes the relationship among various economic sectors and can simulate economic activities and the impact of policies. Since the 1970s, the CGE model has been widely used in evaluating the national economy, trade, environment, finance and taxation, and public policy, and it is one of the main tools of economics and public policy, internationally recognized as a basic tool for analysis by the World Bank and international trade organizations.

An important concept of the CGE model is the "general equilibrium." Its origin can be traced back to 1874, when the Lausanne school leader and French economist Walras, in *Elements of Pure Economics*, first proposed a general equilibrium concept, namely, the "Walras general equilibrium" (Walras and Jaffé 1954). Walras expresses the idea of Adams' "invisible hand" as a set of observable equations, the solution of which is equilibrium under certain conditions.

Another key point of the CGE model is how to realize "computable" and how to solve the two problems. In 1912, the development of the topology of the fixed-point theory provided a powerful tool with which to prove the existence of the equilibrium state of the economy. Wald (1951) first proved the existence of the static equilibrium solution in the complete and imperfect competitive market. However, this is only a special case of the general equilibrium system. Arrow and Debreu (1954) used set theory, topology, and other more abstract tools, refining the idea of Walras, and used the Kakutani fixed-point force that has proved the existence of the equilibrium price in line with the Pareto optimal in economy. This is a landmark achievement in theoretical economics and is of epoch-making significance.

The authors of this chapter are Changxin Liu and Yi Sun.

Z. Wang, *Principles of Regional Science*, New Frontiers in Regional Science: Asian Perspectives 15, DOI 10.1007/978-981-10-5367-2_7

However, because its proof is nonstructural, and cannot tell us how to get a balanced price, the "computable" point is still not achieved.

Johansen (1960) adopted the international trade theory, economic growth theory, and the widespread use of public finance in a special model: a two-part growth model to analyze the various policy changes, the general equilibrium effect. Under the assumption of equilibrium, the ordinary linear differential equations are transformed into linear equations. This is the basic method of qualitative theory of differential equations, which is the basis of nonlinear analysis. Thus, CGE models and computing systems were born. After the pioneering work of Johansen (1960), Scarf (1973) and Shoven and Whalley (1972) published general equilibrium solutions for new research, uniqueness, stability, and optimality; thus, the CGE model method is mature and perfect.

CGE can be extended to a large extent because of the linearization method of Johansen. At present, the method of solving CGE has the solution of linear equations, and also the optimization software package of GAMS software. But either way, it is necessary to linearize the equation, which is the linearization described by Johansen. In essence, GAMS uses the linear programming method as its optimization method. The essence of linear programming is to solve the equations on the vertex of the feasible region. Therefore, it can be said that the current solution to CGE is that first the nonlinear equations are linearized, then the linear equations.

Because the linearization method described by Johansen first takes the natural logarithm of the variable, then the differential, its purpose is to solve the change rate of the CGE variable and then the absolute value of the variable. In the actual simulation analysis, the change rate is more meaningful than the absolute value. Take Eq. (7.1) for example; the result becomes the linear Eq. (7.2). The meaning of the variables after linearization is change rate. This is an approximate treatment, and errors exist. When the variables are small, they can be approximated as equal. However, it is different from the linear approximation in the general mathematical sense. In mathematics, linear approximation is used to make the function near a certain point, and then to do the Taylor expansion.

$$Y_i = A_i K_i^{\alpha_i} L_i^{1-\alpha_i} \tag{7.1}$$

$$y = a + \alpha k + (1 - \alpha) l \tag{7.2}$$

where
$y = d\ln Y, a = d\ln A,\ \ k = d\ln K, l = d\ln L.$

7.1.2 The Basic CGE Model

The CGE model constructed in this book is a closed Chinese dynamic CGE model. In addition, the data utilized in this book are the input–output table of 2007, and thus the dynamic starting year is 2007. The running time interval is 1 year each period.

Production Technology Structure

This book describes the production of technology structure of CGE model that basically uses traditional methods. The total output of each sector is through the intermediate input and production factors in the form of the Leontief composite, which is mainly assumed between intermediate inputs and factors of production that cannot be alternative, although in reality one can buy advanced equipment by investing more capital, thus saving raw materials from immediate use and the necessity of their replacement. However, because the data cannot be obtained, the book is still in the form of Leontief. The production factor is composed of the capital and labor force in the form of the Cobb-Douglas (C-D) production function, and to meet the return on the same scale; at this time, the sum of the elasticity of substitution between labor and capital is 1. The total intermediate input through the various sectors of the product is in the form of Leontief, which means that the products as the use of intermediate inputs cannot be exchanged among each other. These relationships can be expressed intuitively in Fig. 7.1.

The equations associated with production technology are described as Eqs. (7.3), 7.4, 7.5, 7.6, (7.7). Equation (7.3) indicates the relationship between the total input and the total output. Equation (7.4) describes the relationship between total output and added value. Equation (7.5) represents the Leontief relationship between the total and intermediate inputs. Equation (7.6) reflects the relationship between the added value and the production factors, which is in the form of C-D. Equation (7.7) is the relationship between capital and labor force, which ensures the lowest cost of input, which is related to the return on capital and wage rate.

$$TINT_j = \rho_j {}^* QS_j \tag{7.3}$$

$$VA_j = \left(1 - \rho_j\right)^* QS_j, \quad \text{where } \rho_j = \sum_{i=1}^{n} a_{i,j} \tag{7.4}$$

$$IT_{ij} = a_{ij}^* * TINT_j, \quad \text{where } a_{ij}^* = \frac{a_{ij}}{\sum_{i=1}^{n} a_{ij}} \tag{7.5}$$

$$VA_j = A_j K_j^{\alpha_j} L_j^{1-\alpha_j} \tag{7.6}$$

$$\frac{K_j r_j}{L_j w_j} = \frac{\alpha_j}{1 - \alpha_j} \tag{7.7}$$

where

QS_j is the total output of sector j, $TINT_j$ is the total intermediate input of sector j, VA_j is the added value of sector j, IT_{ij} is intermediate demand of sector j for sector i, K_j is the capital demand for sector j, L_j is labor input for sector$j$, r_j is capital return rate of sector$j$, w_j is wage of sector j, and a_{ij} is the input–output coefficient

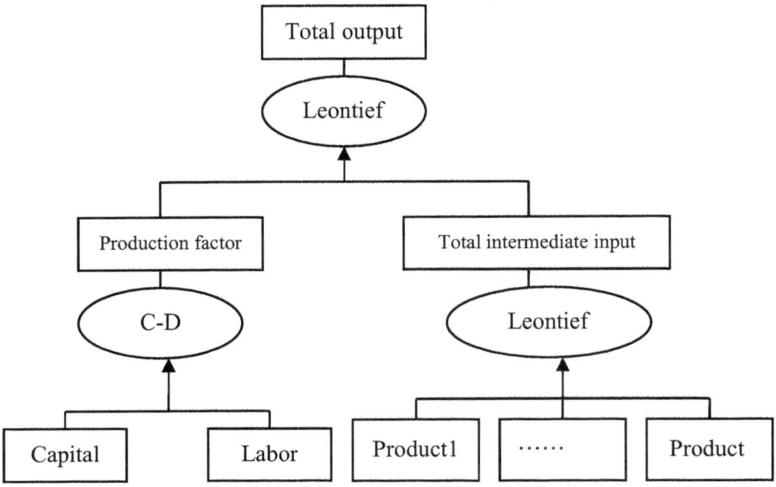

Fig. 7.1 The structure of the production technology of the computable general equilibrium (CGE) model

- Demand structure

Because the model we constructed here is a closed CGE system, when compared with the open CGE system there are no import and export items, or foreign investment, etc. Import demand should be combined into the demand for residents. Therefore, in the closed CGE system, the demand structure comprises five parts: the investment demand, the resident demand, the government demand, the stock demand, and the intermediate input demand (see Eq. (7.8)). Intermediate input demands are described by the Leonitef input–output matrix (see Eq. (7.9)). The demand equation of residents is described by a linear expenditure system (see Eq. (7.10)). The government demand and investment in each sector are allocated to products in accordance with the principle of utility maximization, the utility function uses the C-D form, and the amount of government investment demand is to maximize the utility principle for Eqs. (7.11) and (7.12). For the linear relationship between inventory and total output, see Eq. (7.13).

$$QQ_j = QINT_j + C_j + G_j + I_j + ST_j \tag{7.8}$$

$$QINT_i = \sum_j a_{ij}^* TINT_j \tag{7.9}$$

$$C_j{}^*P_j = P_j{}^*\gamma_j + \beta_j\left(Yu - \sum_{i=1}^{n} P_i\gamma_i\right) \tag{7.10}$$

$$G_j{}^*P_j = \varphi_j{}^*Yg \tag{7.11}$$

$$I_j{}^*P_j = \phi_j{}^*Itot \tag{7.12}$$

$$ST_i = \lambda_i QS_i \tag{7.13}$$

where QQ_j is the total demand of sector j, $QINT_j$ is the total intermediate input demand of sector j, C_j is the resident consumption for sector j, P_j is the product price of sector j, γ_j is the basic resident demand for sector j, Yu is the income of the resident, G_j is the resident consumption of sector j, Yg is the income of government, I_j is the demand of investment of sector j, ST_j is the stock demand of sector j, and $\beta_j, \varphi_j, \phi_j, \lambda_j$ are proportion share parameters.

- Income and savings

The economic entities of CGE are residents, enterprises, and governments. The three have corresponding expenditures, but also have corresponding income and savings. The income of residents and enterprises is always associated with tax. Equation (7.14) indicates that the source of income of the residents is the value of the income tax paid by the workers and the return on capital. Equation The (7.15) indicates that the income tax is proportional to the income of the residents. Equation (7.16) indicates that the government's revenue comes from the indirect taxes on production, the direct tax, and income tax. It also includes the government's capital gains. Equation (7.17) indicates that the indirect tax is derived from the added value of factor production. Equation (7.18) indicates that the income of an enterprise is the direct tax paid by the enterprise. The relationship between direct tax and enterprise income can be expressed by Eq. (7.19). Equations (7.20), (7.21), and (7.22) indicate the distribution relationship among residents, enterprises, and government. Equations (7.23), (7.24), and (7.25) represent the savings of residents, businesses, and governments, respectively. Equation (7.26) means that the gross national product is equal to the sum of the added value of each department.

$$Yu = \sum_i L_i w_i + Ku\_r - Tu \tag{7.14}$$

$$Tu = Yu^* td \tag{7.15}$$

$$Yg = Tu + Te + \sum_j TX_j + Kg\_r \tag{7.16}$$

$$TX_j = tx_j{}^* \left(L_j w_j + K_j r_j\right) \tag{7.17}$$

$$Ye = Ke\_r - Te \tag{7.18}$$

$$Te = Ye^* te \tag{7.19}$$

$$Ke\_r = s_1{}^* \left(\sum_i K_i r_i\right) \tag{7.20}$$

$$Ku\_r = s_2{}^* \left(\sum_i K_i r_i\right) \tag{7.21}$$

$$Kg\_r = s_3{}^*\left(\sum_i K_i r_i\right) \tag{7.22}$$

$$Su = Yu - \sum_i C_i P_i \tag{7.23}$$

$$Se = Ye \tag{7.24}$$

$$Sg = Yg - \sum_i G_i P_i \tag{7.25}$$

$$GDP = \sum_j VA_j{}^* PVA_j \tag{7.26}$$

where

Ku_r, Ke_r, Kg_r represents the capital investment income of residents, enterprises, and government, Tu, td represents the income tax and its rate of residents, Te, te represents the income tax and its rate of enterprises, TX_j represents the indirect income of sector *j*, Ye is the income of enterprises, Su, Se, Sg represents the residents, enterprises, and the government's savings, respectively, GDP is gross domestic product (GDP), r_i, w_i represents the capital return rate and the wage of sector *i*, and s_1, s_2, s_3 represents the allocation coefficients.

- Pricing system

The main factors involved are the price of the elements, the price of the intermediate inputs, the final product price, and the price of the added value (which can be regarded as the total factor price). Equation (7.27) reflects the price of intermediate inputs, which are related to the direct consumption coefficient and the price of the product. Equation (7.28) reflects the relationship between the price of the elements and the price of each factor and the tax rate. Equation (7.29) reflects the relationship between the product price and the composition price and the price of intermediate goods.

$$PTINT_j{}^* TINT_j = \sum_i a_{ij}{}^* QS_j{}^* P_i \tag{7.27}$$

$$PVA_j{}^* VA_j = \left(L_j w_j + K_j r_j\right)^* \left(1 + tx_j\right) \tag{7.28}$$

$$P_j{}^* QS_j = PVA_j{}^* VA_j + PTINT_j{}^* TINT_j \tag{7.29}$$

where

$PINT_j$ is the price of the intermediate inputs, PVA_j is the price of the added value, and tx_j is the product tax rate of sector *j*.

- Market clearing

Market mainly includes the labor market, capital market, product market, and investment savings balance. Equation (7.30) reflects the product market clearing; the total supply of each department is equal to the total demand. Equation (7.31) reflects the clearing of the capital market; the total investment is equal to the total savings. Equation (7.32) and Eq. (7.33) reflect the elements of market clearing. Equation (7.32) is the total capital supply, which equals the total capital demand. Equation (7.33) is the total supply of labor, equal to the total demand of the labor force.

$$QS_j = QQ_j \tag{7.30}$$

$$Itot = Su + Sg + Se - \sum_i ST_i \tag{7.31}$$

$$\sum_i K_i = Ks \tag{7.32}$$

$$\sum_i L_i = Ls \tag{7.33}$$

where Ls represents the total supply of labor force and Ks represents the total capital supply.

- Macroscopic closure

To ensure the uniqueness of the CGE solution, we must make some assumptions on the model. We should eliminate a set of constraints, one of which is the unemployment rate, public expenditure, investment level, and income level. This is the macroscopic closure of CGE. There are four kinds of macroscopic closure: Keynes closure, Jonhansen closure, new Keynes closure, and neoclassical closure. The Keynes closure allows the existence of unemployment, the Jonhansen closure is assumed to maintain a given level of exogenous investment, and the New Keynes closure assumes that the wage is fixed. Neoclassical closure generally assumes that the level of investment is endogenously determined, and that savings determine the amount of investment.

In this book, the closed dynamic CGE model uses a new classical closure. That is, the level of investment is endogenously determined, which is the part of the income of the residents and the government. Capital supply and labor supply are set as exogenous.

- Dynamics

The dynamic mechanism of CGE is usually expressed as the capital recursive. It should be noted that the investment $Is_{i,t}$ and the investment I_i in Eq. (7.8) are not the same. $Is_{i,t}$ is the amount of investment that will be made for the next phase of production, but I_i is the demand for all sectors of the product, and this demand is not used for consumption, but to buy their products for investment production. For example, many sectors will have a demand for equipment manufacturing industry products, and this demand is mainly for the production of investment, that is, not a demand to invest in the department for production. But there is a relationship between the two investments: that is, the total investment demand is equal to the total amount of investment (see Eq. (7.35)).

$$K_{i,t+1} = Is_{i,t} + (1 - \delta_{i,t})K_{i,t} \tag{7.34}$$

$$\sum_i I_{i,t} = \sum_i Is_{i,t} \tag{7.35}$$

where δ is the depreciation rate.

In addition, the labor force growth and technological progress (total factor productivity) should be taken into account. The growth rate of labor force and total factor productivity have a tendency to decrease gradually.

$$\frac{\dot{A}_t}{A_t} = a^* e^{bt} \tag{7.36}$$

$$\frac{\dot{L}_t}{L_t} = l^* e^{mt} \tag{7.37}$$

where $0 < a, l < 1, \quad b, m < 0$.

7.1.3 SAM Table: The Data for CGE

The CGE database is the SAM table. The SAM table is a detailed representation of the accounting system of a social economic system (Round and Pyatt 1985; Roland-Holst et al. 1992). It increases the nonproduction sector based on the input–output table (institutional accounts), such as residents, government, and other parts of the world, between the performance of the production activities, the production elements, and the mechanisms of income, consumption, and capital formation of contact. The input–output table can only show the interdependence between the central sectors of the production account, mainly in the performance of production activities (National Bureau of Statistics 1999). The residents, businesses, and government sources of income are not taken into account. Therefore, compared with the input–output table, the content of SAM is more abundant, reflecting the various economic relationships between the economic and social subjects in a certain period. In short, the input–output table is concerned with the relationship between the production sector, and the SAM table is concerned with the entire national economy, the balance problem. SAM is an extension of the input–output table, but the input–output table is the core of the SAM table.

There are four ways to express national economic accounting: graphic method, balance report, account balance, and matrix form; the matrix form is the advanced form of accounting. The SAM table is the matrix form. The rows in the SAM table correspond to the same column in the corresponding location. Rows represent the income of the accounts; the column shows the expenditure of the account. Income equals expenditure, that is, each account is balanced. The structure form of the SAM table is not unique; the type and quantity of the account settings can be adjusted according to the actual needs. In general, however, the typical SAM accounts of an open economy typically include goods, activities, factors, residents, businesses, government, capital, and other parts of the world. For closed economies, there are no other parts of the world. For the China closed CGE system SAM table, you can use Table 7.1) to reflect the specific form.

You can see that the SAM table can be very clear to express the relationship among the various accounts. The basic relationship of the SAM table is to meet the "total demand = total supply, total output = total investment" and so on. We do not explain here the specific meaning represented in the table, but there are many reference answers (see Wang et al. 2010a, b, c, d, e). It c because of the construction of CGE in economy and refining the Department for the government and enterprises, the transfer payment is not considered; therefore, we do not involve the government transfer payment to the enterprise, and the residents also do not involve the transfer of enterprises for residents to pay.

To make the SAM table, we find the relevant data from the input–output table of China in 2007, the cash flow statement of China in 2007, and the statistical yearbook of China in 2007. The SAM table of China's closed economy is compiled (as shown in Table 7.2). It is easy to find that some of the corresponding rows and columns of the table are not equal, which means the SAM table is not balanced, mainly because there ise a variety of data sources.

To get the balanced SAM table, the table needs to be adjusted (Table 7.2). There are some technical methods for SAM balancing: common RAS (rows and columns, RAS) and cross-entropy method (cross-entropy, CE). In this book, the RAS method is adopted. Adjustment results are shown in Table 7.3. The data of the SAM table itself are very important for building a balanced CGE. In addition, some parameters, such as the transfer payment rate and the proportion of capital allocation, should be estimated according to the SAM table. Therefore, the SAM table is very important for the calculation of CGE.

7.1.4 An Example of Applying CGE Model

In this section, we use the CGE model to analysis the governance of China's volatile organic compound pollution.

Volatile organic compounds are defined by the United States Environmental Protection Agency (US EPA) as "organic chemical compounds whose composition makes it possible for them to evaporate under normal indoor atmospheric conditions of temperature and pressure." Since the photochemical pollution of Los Angeles in the early 1950s, VOCs have been a topic of public concern. Policies are necessary for emissions control given the deleterious impacts of VOCs on the environment and on human health. For instance, VOCs cause stratospheric ozone depletion and tropospheric photochemical ozone formation and contribute to the global greenhouse effect. Additionally, compounds such as toluene, ethylbenzene,

Table 7.1 The SAM table of China

	Commodity	Activity	Labor	Capital	Residents	Enterprise	Government	Investment	
Commodity		Intermediate demand			Consumption		Consumption	Invest demand	Total demand
Activity	Total output								Income of production factor
Labor		The added value from labor							
Capital		The added value from capital							Income
Residents			Labor income	Capital income		Transfer expenditure	Transfer expenditure		Income
Enterprise				Capital income			Transfer expenditure		Income
Government		Indirect tax income		Capital income	Income tax	Direct tax			
Investment					Savings	Savings	Savings		Total savings
	Total supply	Total input	Expenditure of production factor	Expenditure	Expenditure	Expenditure	Total investment		

Table 7.2 The initial SAM table (unit, 10 thousand RMB)

	Commodity	Activity	Labor	Capital	Residents	Enterprise	Government	Iinvestment	
Commodity		5,528,151,509			1,180,730,547		351,909,186	1,109,194,214	8,169,985,457
Activity	8,188,589,620								8,188,589,620
Labor		1,100,473,000							2,275,250,878
Capital		1,174,777,878							
Residents			1,100,473,000	247,415,656					1,347,888,657
Enterprise				804,015,633					804,015,633
Government		385,187,233		123,346,589	20,939,600	290,836,822			820,310,243
Investment					146,218,509	494,574,648	468,401,057		1,109,194,214
	8,188,589,620	8,188,589,620	2,275,250,878		1,347,888,657	785,411,470	820,310,243	1,109,194,214	

Table 7.3 The balanced SAM table of China (unit, 10 thousand RMB)

	Commodity	Activity	Labor	Capital	Residents	Enterprise	Government	Investment	
Commodity		5,531,932,062			1,182,348,400		353,795,674	1,114,980,796	8,183,056,932
Activity	8,183,056,932								8,183,056,932
Labor		1,100,099,251							2,267,727,856
Capital			1,167,628,605						
Residents			1,100,099,251	248,765,426					1,348,864,677
Enterprise				795,335,780					795,335,780
Government		283,397,014		123,527,399	20,856,575	294,485,111			822,266,099
Investment					145,659,702	500,850,669	468,470,425		1,114,980,796
	8,183,056,932	8,183,056,932	2,267,727,856		1,348,864,677	795,335,780	822,266,099	1,114,980,796	

xylene, and trimethyl-benzenes are hazardous to human health because of toxic or carcinogenic properties.

Given the wide range of emissions sources, such as pesticides, paints, building materials, and household products, and China's rapidly growing economy and consumer class, VOC pollution has become a pressing problem. In recent years, attention has been drawn to VOCs in particulate matter (PM) pollution in urban areas such as Beijing–Tianjin and the Pearl River Delta (PRD). VOC pollution impacts about one fourth of China's geography and threatens the health of 600 million of the country's citizens. For example, in Beijing, 9.3% ± 3.0% of the major particulate matter (PM) contains benzene-soluble components. Based on health risk assessment of atmospheric VOCs in urban–rural junctures, the cancer risk index from carcinogenic benzene exposure exceeds the EPA standard.

VOCs have been regulated in developed countries such as the US for several decades, but developing economies such as China have lagged much farther behind, at both national and regional levels. According to Wang and colleagues, China produced more than 6.5 million tons of VOC emissions in 2010, a level that will continue to increase with economic growth. Since May 11, 2010, the Ministry of Environmental Protection (MEP) has classified VOCs as major air pollutants, marking these for significant national measures. In 2013, China published the "Action Plan of Atmospheric Pollution Control," namely "Ten policies," to control the haze, and requires the 2017 prefectural level of PM 10 concentration be reduced more than 10% to the 2012 level. The "Atmospherics Pollution Prevention Law" was amended in November 2013, and the country wishes to establish a joint prevention and control mechanism for regional atmosphere pollution. The law requires managing VOCs from the pollution sources and has tightened sanctions against enterprises that break the law. China's primary policy instrument for VOC abatement is pollution charges; in recent years, however, there have also been calls for the implementation of environmental taxes. The objective of this study is to compare pollution charges and environmental taxes in terms of their economic and environmental impacts, with direct implications for the future regulation of VOC emissions in China.

A pollution charge is a policy requiring enterprises to pay the cost of collection, treatment, and disposal of sewage, following the "Polluter Pays Principle." An environmental tax is designed to internalize environmental costs and motivate the public to achieve sustainability, with the financial burden borne by the clientele of a polluting industry: this follows the "Consumer Pays Principle." Engineering reduction methods refer to the traditional treatment techniques by which VOC emissions are reduced through thermal combustion, catalytic oxidation, biofiltration, low-temperature plasma decomposition, photocatalytic degradation, adsorption/absorption, and their composite techniques. Used in different industry sectors, these technologies have different reduction rates. Their selection is based on both feasibility and economic costs.

In China, pollution charges have existed since 1982, but there have been no environmental taxes until the present time. Before 1993, plants were only required to pay fees when effluent discharges exceeded the legal standard. In 2003, the State

Council issued "Regulations on the levy use of pollutant discharge fees," and started to manage pollutants by controlling total emission quotas. Enterprises pay uniform rates for within-standard emissions and higher but de-escalating rates for above-standard emissions. When the pollutant level is lower than the corresponding standard, plants pay a within-standard fee for the total amount; otherwise, they pay the above-standard fee. Billions of RMB have been collected from industries for air, solid waste, water, and noise pollution, which fees were then used to fund environmental projects.

In recent years, however, it has been claimed that environmental taxes are a more optimal instrument for regulating pollution, as it can lead to a "double dividend:" better environmental standards, fewer distortionary taxes, and welfare improvement. Thus, although pollution charges have had a significant role in VOC reduction in China over recent decades, the introduction of environmental taxes may be more consonant with the demands of China's increasingly industrialized and market-oriented economy.

We use a computable general equilibrium (CGE) model that was developed to simulate the macroeconomic effects of public policies in both developed and developing countries. Such models are widely used to analyze the effects of government action in fields such as finance, price, tax, and international trade. Developed in the 1960s, these models have also become a popular tool for determining relationships between economic activity and environment outcomes. In particular, CGE models have been adapted to explore the effectiveness of pollution control policies since the 1980s. Pollutant emissions can be estimated by using fixed pollution coefficients. However, it is still difficult to completely quantify dynamic economy–environment relationships, and pollution inventory data are often lacking, especially for developing countries. Therefore, this kind of model application is still relatively new in China.

The purpose of this chapter is to compare the different impacts of pollution charge and environmental tax on VOC reduction in China. The pollution charge will impact the production costs whereas the tax will affect the goods price directly. Our first step is to ascertain the amount of emissions reduction through the application of pollution charges when the charge is equal to the average cost of engineering reduction methods. Then, we calculate the ratio of the environmental tax fee when it has the same reduced emissions as the pollution charge (see Eqs. 7.38 and 7.39). Finally, we compare their impacts on the macro-economy using GDP, sector output, and other major assessment indicators.

The models used for VOC reduction assessment are built upon previous CGE models developed by Zhu and colleagues. These models provide a description of the Chinese economy in 2007, when the country's official input–output table became available in published statistics yearbooks. In line with the data from the input–output table, 42 sectors are classified in the model, numbered as S_i ($0 < i < 43$; i is integer). We adapted and incorporated the national economy and VOC emissions inventory in our analysis. Figure 7.2 outlines the basic structure of the model, which explicates the interactions between the environment and the economy. In the model, the sector aggregate supply is equal to the aggregate

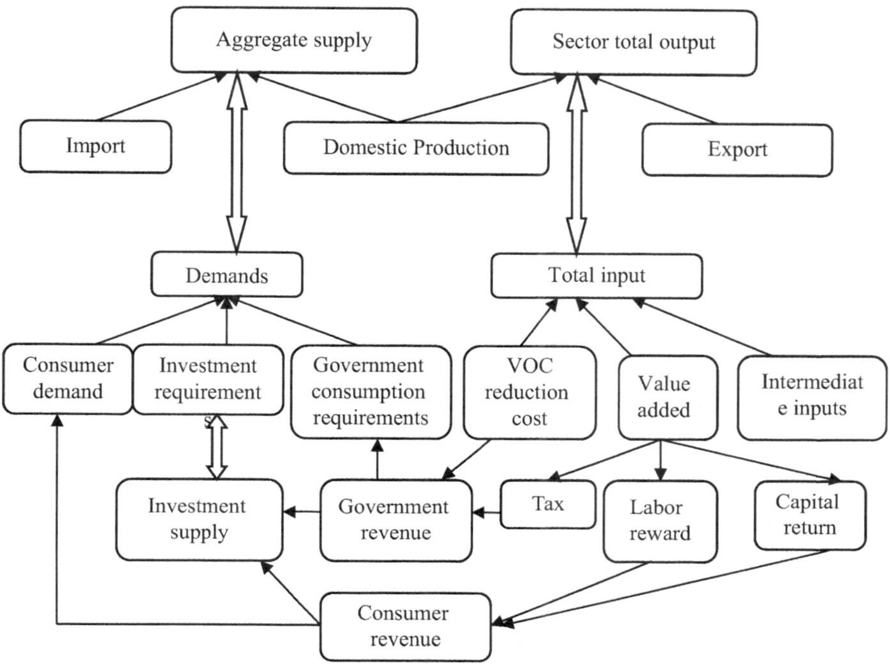

Fig. 7.2 Basic structure of the CGE model system

demand, and the total output is equal to the total input. Moreover, there are three different parts of sector input: value added, intermediate input, and VOC reduction cost. The revenue of residents and the government is utilized for investment and consumption. Additional details of the model are described in Tables 7.4 and 7.5.

Next, we show the differences between pollution charges and environmental taxes when they are simulated in the CGE model. Their microeconomic mechanisms are demonstrated in Fig. 7.2. In the left sub-figure of Fig. 7.3, p^a and p^* are supply price with and without pollution charges, respectively. In the right sub-figure, supply price and consumption price are indicated as p^s and p^d, respectively. The equilibrium state can be reached when the tax is equal to the difference between p^s and p^d. Production will decline from increases in charges and taxes if other conditions remain the same. Enterprises can continue to gain benefits when the marginal cost of production reaches the market price. The reasons why the environmental tax is based on the indirect tax have been analyzed in a previous study [13]. It is not easy to directly compare pollution charges and environmental taxes in a complex and dynamic economy, as they are influenced by the industrial supply chain as well as myriad market forces. However, CGE models help avoid this disadvantage and can clearly describe the effects of policy implementation (Table 7.6).

In the CGE model, pollution charges and environmental taxes can be expressed as the following equations, Eqs. (7.38) and (7.39). Formulation (7.38) represents the

Table 7.4 Sector names and serial numbers

Serial number	Sectors	Serial number	Sectors
S1	Agriculture, forestry, animal husbandry, and fisheries	S22	Scrap and waste
S2	Mining and washing of coal	S23	Production and supply of electric power and heat power
S3	Extraction of petroleum and natural gas	S24	Production and distribution of gas
S4	Mining of metal ores	S25	Production and distribution of water
S5	Mining and processing of nonmetal ores and other ores	S26	Construction
S6	Manufacture of foods and tobacco	S27	Traffic, transport, and storage
S7	Manufacture of textiles	S28	Post
S8	Footwear, caps, leather, fur, feathers (down) and its products	S29	Information transmission, computer services, and software
S9	Processing of timber and manufacture of furniture	S30	Wholesale and retail trades
S10	Papermaking, printing, and manufacture of articles for culture, education, and sports activities	S31	Hotels and catering services
S11	Processing of petroleum, coking, processing of nuclear fuel	S32	Financial intermediation
S12	Chemical products	S33	Real estate
S13	Manufacture of nonmetallic mineral products	S34	Leasing and business services
S14	Smelting and rolling of metals	S35	Research and experimental development
S15	Manufacture of metal products, smelting and pressing of ferrous metals	S36	Comprehensive technical services
S16	Manufacture of general-purpose and special-purpose machinery	S37	Management of water conservancy, environment, and public facilities
S17	Manufacture of transport equipment	S38	Services to households and other services
S18	Manufacture of electrical machinery and equipment	S39	Education
S19	Manufacture of communication equipment, computers, and other electronic equipment	S40	Health, social security, and social welfare
S20	Manufacture of measuring instruments and machinery for cultural activity and office work	S41	Culture, sports, and entertainment
S21	Manufacture of artwork, other manufacturing	S42	Public management and social organization

Table 7.5 Equations of the CGE model system

Equations	Basic description
$Q_i = A_i^q\left(\delta_{i1}M_i^{\rho_i} + \delta_{i2}D_i^{\rho_i}\right)^{1/\rho_i}$	Aggregate demand equation
$\frac{PM_i{}^*M_i}{PD_i{}^*D_i} = \frac{\delta_{i1}M_i^{\rho_i}}{\delta_{i2}D_i^{\rho_i}}$	Equation between domestic demand and import demand
$P_iQ_i = PM_i{}^*M_i + PD_i{}^*D_i$	Price of composite commodity
$Q_i = \sum_j a'_{ij}IT_j + C_i + G_i + I_i$	Equilibrium of demands
$C_i = \frac{\varphi_i Yu}{P_i(1+t_i)}$	Residents' demand
$G_i = \frac{\beta_i Yg}{P_i(1+t_i)}$	Government demand
$I_i = \frac{\gamma_i I_{\text{tot}}}{P_i(1+t_i)}$	Investment demand
$X_i = D_i + E_i$	Total output equation
$E_i = \omega_i \cdot \left(\frac{PWE_i}{PE_i}\right)^{\eta_i}$	Export demand equation
$PX_i{}^*X_i = PD_i{}^*D_i + PE_i{}^*E_i$	Output price equation
$X_i = A_i\left(\delta_{i1}VA_i{}^{\rho 2_i} + \delta_{i2}IT_i^{\rho 2_i}\right)^{1/\rho 2_i}$	Technology equation
$\frac{PVA_i{}^*VA_i}{PIT_i{}^*IT_i} = \frac{\delta_{i1}VA_i{}^{\rho 2_i}}{\delta_{i2}IT_i^{\rho 2_i}}$	Equation of intermediate inputs, labor inputs, and capital inputs
$PIT_i = \sum_j a_{ij}{}^*P_j$	Intermediate production price
$VA_i = A_iK_i^{\alpha_i}L_i^{1-\alpha_i}$	Technology equation of added value
$K_ir_i = \frac{\alpha_i{}^*L_iw_i}{1-\alpha_i}$	Equation of investment capital and labor
$PVA_i{}^*VA_i = K_ir_i + L_iw_i$	Price of added value
$V_i = X_i{}^*Ef_i$	Volatile organic compound (VOC) emissions
$PX_i{}^*X_i = PVA_i{}^*VA_i + PIT_i{}^*IT_i + f_i{}^*V_i + TX_i$	Cost equation with VOC abatement cost
$VOC = \sum_i V_i$	Total VOC emissions
$Tx_i = (C_i + G_i + I_i)P_i{}^*t_i$	Tax equation
$Yu = \sum_i (K_ir_i + L_iw_i)$	Residents' income
$Yg = \sum_i Tx_i$	Government revenue
$Su = Yu - \sum_i C_iP_i(1+t_i)$	Household savings
$Sg = Yg - \sum_i G_iP_i(1+t_i)$	Government savings
$Su + Sg = I_{tot}$	Capital clearing equation

pollution charges that would increase the production cost and affect the total product amount and the production price. Formulation (7.39) represents the tax that would influence consumers' price and the amount of the product, which means the tax would be passed on to consumers directly. Thus, the fee would increase the price of production and the tax would increase the price to consumers. Although the

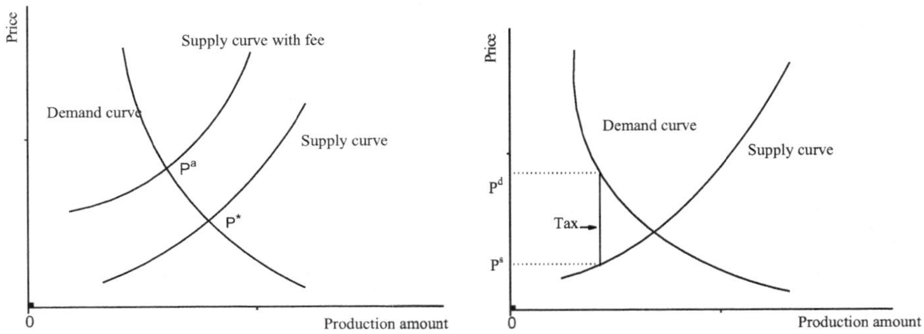

Fig. 7.3 Microeconomic mechanisms of pollution charges and environmental taxes

Table 7.6 Basic description of model parameters

Variables	Basic distribution	Variables	Basic distribution
a	Input–output coefficient	*PIT*	Prices of intermediate goods
C	Household consumption	*PM*	Price of imported goods
C_{tot}	Total consumption of household	*PVA*	Price of added value
D	Domestic supply	*PWE*	World prices of export goods
E	Export supply	*PWM*	World prices of imported goods
f	Pollution charge fees of VOCs	*PX*	Prices of suppliers
G	Government consumption demand	*Q*	Aggregate demand
GDP	Gross domestic product	*r*	Return of capital
G_{tot}	Total government consumption	*Sg*	Government savings
I	Investment demand	*Su*	Household savings
IT	Intermediate import	*t*	Tax rate
I_{tot}	Total investment	*TX*	Tax
K	Capital	*VA*	The added value
L	Labor force	VOC	VOC emissions
M	Import demand	*w*	Wage rate
P	Demand for composite products	*X*	Total output
PD	Price of domestic production	*Yg*	Income of government
PE	Prices of export products	*Yu*	Income of individuals
		EF	Emission factors without control measures

final result is that prices would go up and production level would go down, there is a different effect on consumers and enterprises.

$$PX_i{}^*X_i = PVA_i{}^*VA_i + PIT_i{}^*IT_i + f_i{}^*V_i \tag{7.38}$$

$$TX_i = (C_i + G_i + I_i)P_i{}^*t_i \tag{7.39}$$

where PX_i is the prices of suppliers of sector i, X_i is the output, VA_i is the added value, PVA_i is the price of added value, PIT_i is the price of intermediate input, IT_i is the quantity of the intermediate input, f_i is the charging fee of VOC emissions/ton in Department i, V_i is the VOC emissions amount in Department i, TX_i is the total tax in Department i, and P_i is the price of domestic goods. C_i, G_i and I_i refer to household consumption, government consumption, and investment demands, respectively; t_i is the tax ratio in Department i.

When China's VOC emissions in 2007 were evaluated, 26.4 million tonnes of anthropogenic VOCs were produced by all 42 economic sectors (details can be found in Fig. 7.4) [29]. Both pollution charges and environmental taxes were imposed on 10 industry sectors, whose emissions in 2007 were each greater than 0.5 million tonnes. These sectors include manufacture of foods and tobacco (S6), manufacture of textile (S7), manufacture of textile wearing apparel, footwear, caps, leather, fur, feather (down) and its products (S8), processing of timbers and manufacture of furniture (S9), papermaking, printing, and manufacture of articles for culture, education, and sports activities (S10), processing of petroleum, coking, processing of nuclear fuel (S11), chemical products (S12), manufacture of transport equipment (S17), production and supply of electric power, and heat power (S23) and construction (S26). Both pollution charges and environmental taxes are imposed on these sectors.

According to the analysis of data from 2007, marginal cost equals the average engineering reduction cost, which is 20,000 RMB/t VOCs. When the pollution charge and the average cost of engineering reduction measures have the same value, it is time for enterprises to choose between engineering control and punishment. In this situation, 2.6% GDP and 520,000 tonnes of VOCs are reduced, accounting for 2.2% of total emissions. This policy generates an economic loss of 1,330,000 RMB/t VOCs. The environmental tax has to be 8.9% higher than the current indirect tax level (17% value-added tax) to achieve a similar impact. It results in a 1.38% reduction of GDP and a loss of 690,000 RMB/t in VOCs. Although the emissions inventory was updated and imposed measures were changed, the result is consistent with a previous study [13]. In considering economic development, for the same level of VOC reduction, an environmental tax causes a smaller decrease in GDP. The average cost of engineering measures was 20,000 RMB/t in 2007, and the cost decreases with increasing scale of project application.

It is necessary to consider how policies affect each of the different sectors. First, environmental taxes and pollution charges exert entirely different impacts on VOC reduction. As a consequence of the pollution charge, emissions are reduced most in the construction sector (S26), followed by the manufacture of general-purpose and special-purpose machinery (S16), manufacture of transport equipment (S17), processing of timber and manufacture of furniture (S9), and manufacture of metal products smelting and pressing of ferrous metals (S15), with decreases,

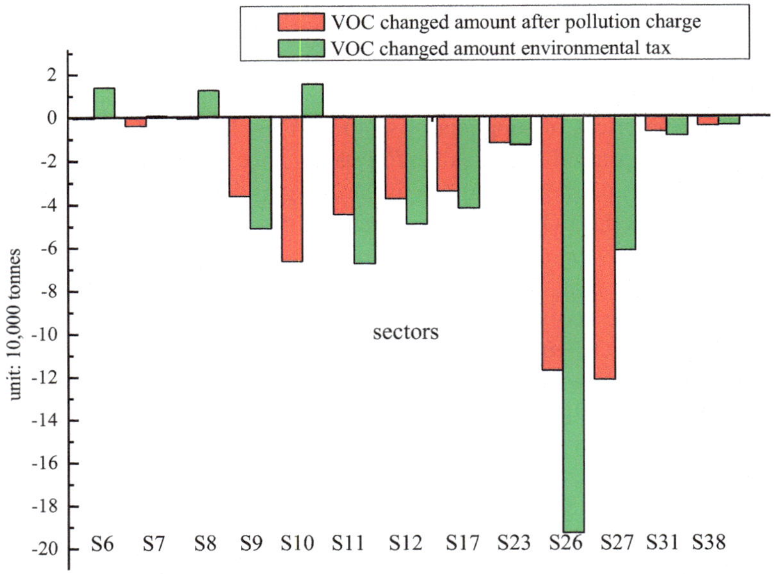

Fig. 7.4 VOC emissions reduction by pollution charge and environmental tax in each sector

respectively, of 117,325 tonnes (8.69%), 9,585 tonnes (4.79%), 34,000 tonnes (3.95%), 36,207 tonnes (3.38%), and 6,568 tonnes (3.28%). The environmental tax had the greatest impact on the construction sector (S26), followed by the manufacture of general-purpose and special-purpose machinery (S16), manufacture of metal products, smelting, and pressing of ferrous metals (S15), manufacture of electrical machinery and equipment (S18), and manufacture of artwork and other manufacture (S21). These sectors witnessed, respectively, emissions decreases of 192,805 tonnes (14.28%), 19,525 tonnes (9.76%), 15,299 tonnes (7.64%), 17,239 tonnes (6.63%), and 6,699 tonnes (5.98%). Other sectors, namely papermaking, printing, and manufacture of articles for culture, education, and sports activities (S10), did not see high reductions.

There is another negative impact. If an environmental tax is implemented, VOC emissions will actually increase in several sectors, such as the manufacture of foods and tobacco (S6), because when the policy is implemented across the entire country, products in sectors such as S6 cannot be easily replaced by imports and the price of products increases because of the tax. At the same time, the output of domestic products in these sectors has a lower elasticity, and the increase of government spending offsets the reduction of household consumption. Consequently, VOC emissions will increase. These two policies also impact the macroeconomy in terms of total output and value-added products (Figs. 7.5 and 7.6). Our study shows that the imposition of a pollution charge can negatively impact the output of all 42 sectors. It directly diminishes the output of the construction (S26) and the manufacture of general-purpose and special-purpose machinery (S16) sectors, causing respective declines of −8.69% and −5.43%. Changes in output

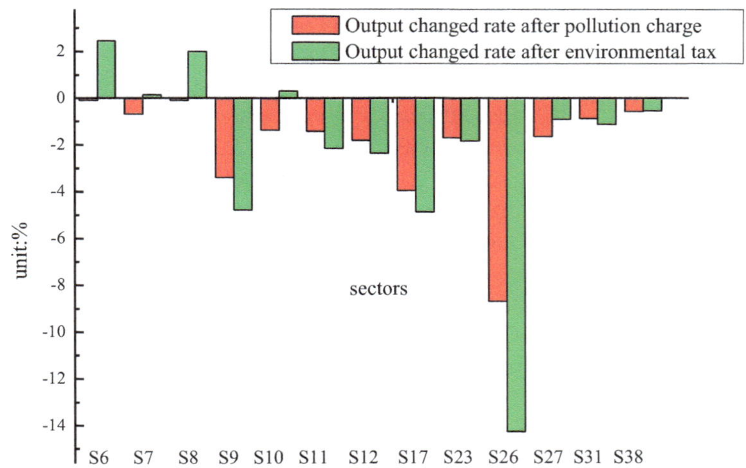

Fig. 7.5 Output changes after the implementation of pollution charges and environmental taxes

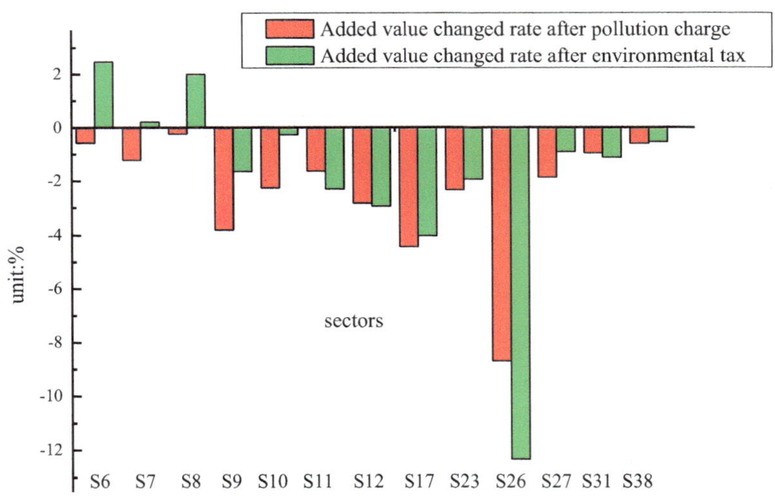

Fig. 7.6 Value-added changes rate after the implementation of pollution charges and environmental taxes

caused by pollution charge implementation in other sectors range from 0% to −3.4%. In terms of output, an environmental tax most affects S26 and S16, which change by −14.28% and −9.76%, respectively. Changes in other sectors vary from 2.5% to −7.65%. These findings indicate that environmental taxes can lead to more notable changes in the output in sub-sectors than pollution charges. Moreover, changes in the total output of each sector, by both pollution charge and environmental tax, have no relationship with shifts in total emissions.

Changes in value-added after the implementation of the two polices are similar to the changes in output. Pollution charges have negative impacts across the board, resulting in shifts of −9% to 0%. The most significantly impacted are S26 and S16, at −8.69% and −5.43%, respectively. Environmental taxes cause the greatest change in the construction sector, with a rate of −12.33%. The sectors of manufacture of metal products, smelting, and pressing of ferrous metals (S15) and manufacture of general-purpose and special-purpose machinery (S16) are the second and third most affected, changing by −8.45% and −8.02%, respectively.

Changes to output and value-added in most subsectors are consistent, meaning when the output rises, the value-added in those subsectors also increases. However, in the papermaking, printing, and manufacture of articles for culture, education, and sports activities sector (S10), the output increases by 0.32% whereas the value-added declines by 0.28%. The major reason is that, in this sector, intermediate consumption and production prices increase after the implementation of policies. Additionally, pollution charges have an even more negative impact on the economy than environmental taxes. Changes in output and value-added in each subsector resulting from the implementation of a pollution charge are therefore negative. Consequently, compared with a pollution charge, an environmental tax exerts a greater influence on output and value-added.

The construction sector, one of the largest and most emissions intensive in China, is strongly affected by upstream sectors, such as the iron and steel and cement sectors. Operations in these sectors turn basic raw materials into intermediary products, such as reinforced concrete and cement, which are finally used in buildings. When the prices of products in these sectors increase, construction sector output will decrease, resulting in the observed changes of output and value-added. Sectors such as manufacture of general-purpose and special-purpose machinery are much more readily influenced by the two policies than other sectors, mainly because products from these sectors lack domestic and international substitutes.

Industrial enterprises are central to China's economy, so it is instructive to look at the project impacts on these sectors. The two key indicators are industry burden and economic loss. Industry burden refers to the initial cost of a policy, as enterprises must incur additional costs to control their pollutants. The burden of a pollution charge is equal to the charge multiplied by the amount of VOCs emitted. The burden of an environmental tax is calculated as the tax rate times the value-added in each sector. Industry burden from a pollution charge is directly decided by the amount of VOC emissions when the charge is equal to the average reduction cost of engineering measures. The more VOCs are emitted, the higher the primary costs that enterprises must pay. The top five sectors impacted by pollution charge are the traffic, transport, and storage sector (S27), papermaking, printing, and manufacture of articles for culture, education, and sports activities (S10), processing of petroleum, coking, processing of nuclear fuel (S11), chemical products (S12), and construction (S26) sectors, with respective burdens of 145.17 billion RMB, 95.1 billion RMB, 61.9 billion RMB, 40.85 billion RMB, and 24.66 billion RMB, accounting for 4.58%, 6.37%, 2.94%, 0.66%, and 0.39% of their sector total output, respectively.

However, an environmental tax can impact both VOC-emitting and non-emitting sectors when the achieved targets are the same as a pollution charge. There is no direct and obvious relationship between emissions reduction and industrial burden caused by the implementation of an environmental tax. Sectors with low emissions, such as the post (S28), manufacture of textile wearing apparel, footwear, caps, leather, fur, feather (down) and its products (S8), and culture, sports, and entertainment (S41) sectors, are negligibly affected by an environmental tax. However, sectors such as extraction of petroleum and natural gas (S3) shoulder a substantial burden from increased taxation.

The top three most affected sectors are those of smelting and rolling of metals (S14), manufacture of general-purpose and special-purpose machinery (S16), and construction (S26). Their corresponding costs are 56.47 billion RMB, 42.3 billion RMB, and 35.3 billion RMB, representing, respectively, 0.93%, 1.07%, and 0.56% of total sector output. The manufacture of general-purpose and special-purpose machinery (S16) and smelting and rolling of metals (S14) sectors produced only 200,000 tonnes and 138,300 tonnes VOCs in 2007, respectively. Their burdens are relatively higher than those of other sectors, whose average loss is only 4.78 billion RMB. In addition, although there are significant VOC emissions from sectors such as the hotels and catering service (S31) and residential consumption (S38), their industrial burdens from an environmental tax are almost zero, mainly because after policies are implemented, their sector outputs decline. In terms of changes to value-added, or "sector economic loss," pollution charges have a negative impact on almost all sectors, ranging from 126.2 billion RMB to 0 RMB. The most losses come from, in declining order, the construction (S26), agriculture, forestry, animal husbandry, and fishery (S1), and the smelting and rolling of metals (S14) sectors. Environmental taxes exert the largest impact on the construction sector, with a loss of 178.9 billion RMB. They also have impacts on other sectors, ranging from −73.1 billion RMB to 25.1 billion RMB. These data show that the implementation of an environmental tax more readily internalizes economic loss and distributes the costs more evenly, resulting in less fluctuation of the macro-economy.

The differences between the economic impacts caused by pollution charges and environmental taxes depend on implementation methods and economic mechanisms. On the one hand, the prices of domestic products increase when a pollution charge is imposed, and the products will be replaced by imported substitutes until a new equilibrium is reached. On the other hand, there is virtually no substitution of domestic products when an environmental tax is implemented, as it is collected from both domestic and imported goods. Considering the current level of management, we propose that a more optimal regulation of VOCs in China requires environmental taxes to be imposed on the ten sectors producing the highest emissions. Additional research is needed to expand that conclusion to other parts of the economy.

We analyzed the differing economic impacts of pollution charges and environmental taxes when the same level of VOC emissions reduction is achieved. About 2.2% of VOC emissions can be abated, from the 2007 level, when a pollution charge is equal to the average cost of engineering measures. For environmental

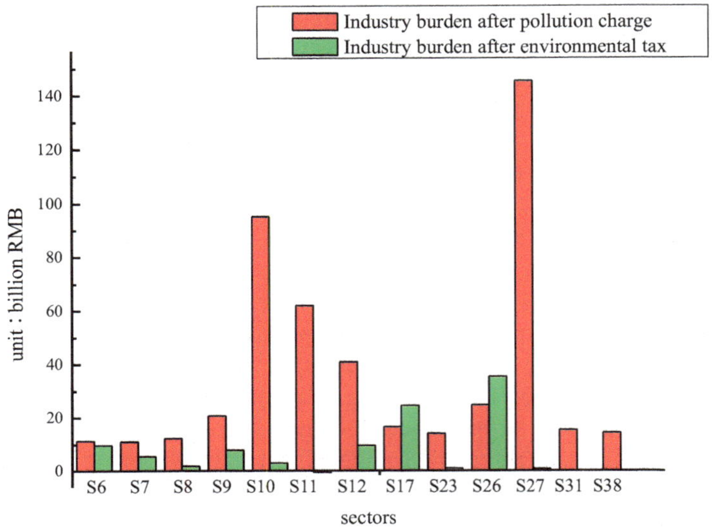

Fig. 7.7 Industry burden in each sector after the implementation of pollution charges and environmental taxes

taxes, an increase of 8.9% over current indirect tax levels needs to be collected from the ten highest VOC-emitting sectors. The impacts on the macro-economy, sector output and value-added, and industrial burden and economic loss were also studied. Based on our analysis, we suggest that in the future, China should favor an environmental tax over pollution charges in VOC regulation. However, we do not make this recommendation without qualification: an environmental tax can also induce a fairness problem among sectors, so policymakers should also carefully consider supplementary measures such as subsidies. Although fiscal and regulatory measures can significantly improve VOC regulation at the national level, emissions can still increase in certain sectors. This circumstance implies that engineering measures are also needed as supplements (Fig. 7.7, 7.8 and 7.9).

7.2 CGE Analysis of Multi-regional Economy

7.2.1 A Regional Computable General Equilibrium Model

There are many economies in the real economic world, forming a multi-economy system or multi-region economy. Some economists represented by Krugman put forward the new economic geography in the early 1990s. Although they did not emphasize multi-region economy, one large progress step of the new economic geography is to replace the traditional single-economy analysis with multi-region economy analysis. Single-economy analysis is difficult to use in multiple

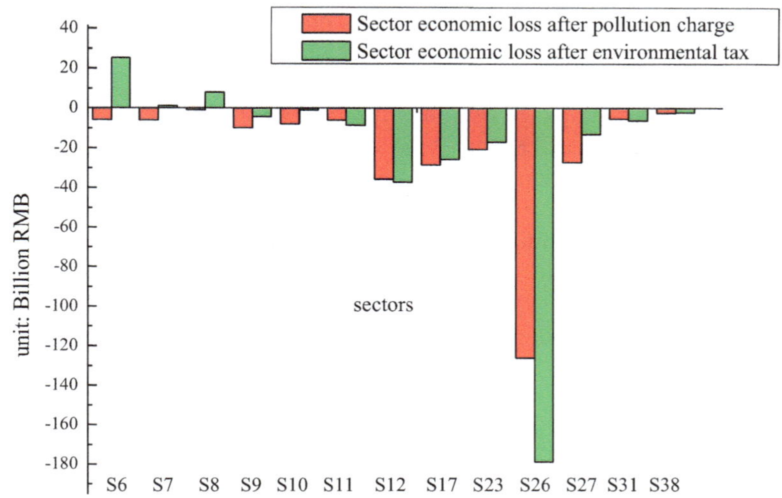

Fig. 7.8 Economic losses in each sector after the implementation of pollution charges and environmental taxes

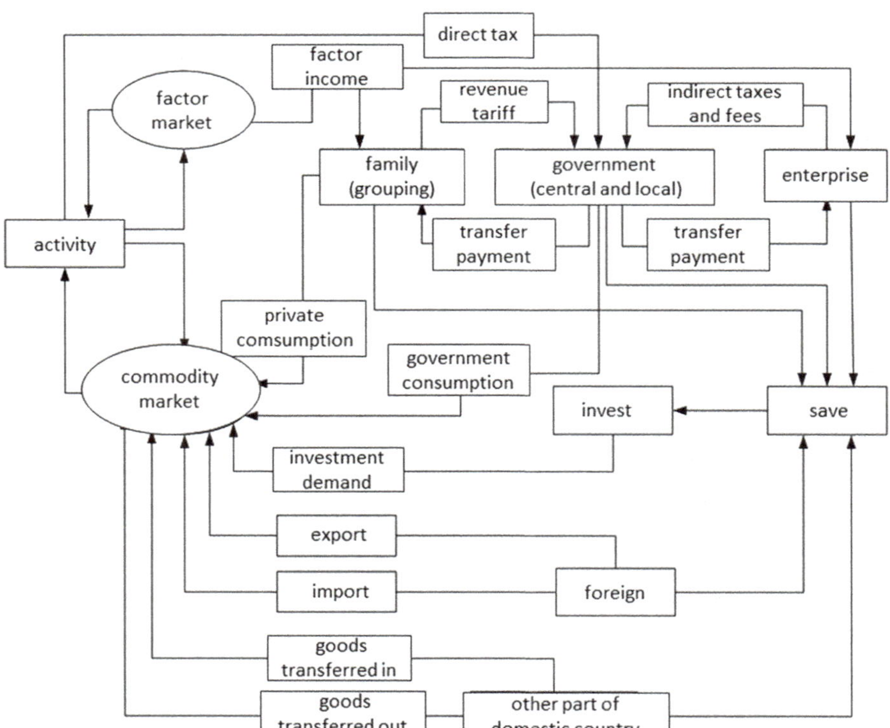

Fig. 7.9 Structure of a single region in the multi-region CGE model of social equity

economies. It is accepted that stimulating consumption could bring economic growth, but in response to the limit of trade and resource supply it may stimulate the consumption of Country A and drive the economy of Country B. For example, stimulating China's computer consumption in 1990s finally promoted America's CPU production growth and Korean's RAM production growth, whereas China was only benefited in production growth of computer cases. Therefore, multi-region economy analysis and policy simulation are necessary. The region described here is an independent economy rather than always a country. For instance, Changsha City, Zhuzhou City, Yueyang City, Shaoyang City, and Changde City are all independent economies and they form the Hunan multi-region economy. According to the analysis perspective, specific places are treated as a multi-region economy in the model.

7.2.2 Multi-regional Computable General Equilibrium Model

General equilibrium is an important analysis concept in economics. In 1874, Walras, a French economist and the leader of the Lausanne school, put forward the concept of general equilibrium for the first time. He thought that an economic system is an integral whole in which there is an equilibrium relationship between suppliers and demanders; otherwise, it will cause excess supply or demand surplus and finally result in economic crisis. As a general rule, an economic system will maintain a balanced market by changing market elements through price adjustment. In other words, investigating market equilibrium and aggregate equilibrium in an economic system, so that the price change caused by supply–demand imbalance under certain conditions can make the supply–demand balanced again, is the basis of economic analysis.

Computable general equilibrium (CGE), also known as application general equilibrium (AGE), uses the general equilibrium method. It makes the complicated general equilibrium economic analysis computable and thereby gives the general equilibrium analysis practical significance. As a model method, CGE pays attention to economic variables of one economy and constructs an interrelationship system based on the general equilibrium of economics, whereas policy provides impact to this system. Policy simulation analyzes policy correlation through these fundamental elements that reflect exogenous variables or system parameters. In methodology, CGE can be viewed as a bottom-up comprehensive integrated method. It depicts industrial departments of the economy differently and then represents them as measurable economic characteristics in considering interaction and interrelationship among economic variables according to the general equilibrium way. It involves many economic departments and has tedious computations. General equilibrium considers correlations of all departments, and that involving some departments is called local equilibrium. Both general equilibrium and local equilibrium may require heavy computation loads. Fortunately, general equilibrium analysis often views economic relationships as linear ones, which simplifies the computation significantly. This approach is feasible because general equilibrium often focuses on equilibrium in the economic system, but changes of the economic system caused by surrounding small policy interference, exogenous variable interference, or system parameter changes could be approximated through linear analysis.

With respect to the theoretical problems of general equilibrium, Walras first expressed the "invisible hand" of Adam Smith as a group of equations and tried to prove the existence of the general equilibrium solution through equations and a number of unknowns. However, this proof was overthrown mathematically. Pareto and Cassel et al. gave a more systematic description to the general equilibrium theory in the early 1990s. The general equilibrium theory achieved remarkable breakthroughs in the 1930s. In the 1950s, economists found that proof about the existence of equilibrium price is not reliable. Arrow (1951) and Arrow and Debreu (1954) refined Walras's idea by using more abstract mathematical tools (e.g., set theory and topology) and proved through the Kakutani fixed-point theorem that there is an equilibrium price conforming to Pareto optimality in limited economy. This is a milestone achievement in economics and has epoch-making significance for economic studies. Arrow and Debreu won the Nobel Economics Price in 1972 and 1983 because of their contributions to the general equilibrium.

Although Arrow (1951) and Arrow and Debreu (1954) achieved great progress in the general equilibrium theory, their proof is unstructured and could not tell us how to obtain the equilibrium price. Therefore, the equilibrium price was still not computable. The first accepted CGE model in the world was developed by Johansen, a Norwegian scholar. To illustrate how to calculate the general equilibrium effect of policy changes, Johansen (1960) employed the international trade theory, the theory of economic growth, and the two-part growth model (an unique general equilibrium model that is widely used in public finance) to analyze the effect of various policy changes. To solve the model, he converted the nonlinear general equilibrium equations into linear ones through logarithmic differentiation under assumed equilibrium. Later, some leading economists in the general equilibrium field further developed and perfected the existence, uniqueness, optimality, and stability of the general equilibrium solution. Scarf (1967, 1973) and Shoven and Whalley (1972, 1974) are leading scholars concerning CGE model application. Based on Lemek's multi-user Nash equilibrium algorithm with limited strategies, Scarf (1967) found an integral convergence algorithm to calculate the fixed point, which makes equilibrium price computable. CGE has been the evaluation standard of practicability of the real economic system in general equilibrium theory since then. In the early 1990s, the CGE model was determined as one of the fields of practical economics, and articles began to appear in some top periodicals and famous publications, such as those by Shoven and Whalley (1974), Pereira and Shoven (1988), Robinson (1991), Bandara (1991), and Bergman (1990). Many books and textbooks about the structure and applications of the CGE model have been published, including those by Johansen (1960), Dixon et al. (1982), Adelman and Robinson (1978), Keller (1980), Harris and Cox (1984), Whalley (1985), Dervis et al. (1982), Shoven and Whalley (1992), and Dixon (1992).

CGE analysis is different from general equilibrium theory: the former is application, whereas the latter is theory. For example, the CGE model eliminates the hypothesis about the complete competitiveness necessity and introduces policy intervention, so that it is now applicable to a multi-country mixed economy (Wu and Wang 2004): it makes the general equilibrium theory closer to economic reality. The CGE model assimilates the advantages of many methods such as input–output and linear planning, and reflects economic ties among departments. In contrast to the input–output model, which neglects market effect, it connects factor market and product market organically through price signals, thus enabling reflection of the interaction of market mechanisms and highlighting of economic ties among different departments.

The CGE model has some outstanding characteristics. (1) The market equilibrium hypothesis: supply–demand behaviors of different economic subjects could determine least price of some commodities and elements. (2) Explicit goal of different economic subjects, such as utility maximization for households, and maximum profit or minimum cost for manufacturers. Through these optimization hypotheses, the role of commodity and element price in decision making of household consumption and manufacturer production could be reflected. (3) Computable characteristic: it can obtain a numerical solution.

The CGE model is widely used in various research fields because of its closeness to economic theory and the diversity of involved economic subjects. After more than 40 years of development, the CGE model is mature enough to analyze the effects of economic policy changes on the economy, including pricing policy, fiscal and tax policy, public consumption policy, climate change policy, energy policy, environmental policy, tariff and exchange rate, economic growth, international trade, technological progress, and salary and income distribution.

7.2.3 General Structure of the CGE

The CGE, a large policy simulation model, possesses a basic model structure. It is generally acknowledged that the CGE model is composed of the following six parts.

1. Production Behavior

In this part, the CGE model mainly describes behaviors and optimization conditions of commodity and element manufacturers, including the production equation and constraint equation of manufacturers, the supply equation of production factors, and the optimization condition equation. The equation that depicts the production behavior mainly illustrates product supply of manufacturers. This equation generally can be divided into two types.

The first type is a descriptive equation. It mainly describes the relationship between production input and output as well as the relationship between intermediate input and output. Producer behavior could be described by the Cobb–Douglas production function, the constant elasticity substitution (CES) production function, and two-layer or multi-layer nested CES production function. Production function can use two traditional production factors—labor force and capital, or use various production factors—labor force, capital, land, and energy sources. Moreover, the labor force could be divided into different groups according to technological level, income, and educational background. In this way, different production functions could be chosen for different research problems to highlight characteristics of the research problem. The intermediate input–output relationship could be described and analyzed with a Leontief input–output matrix.

The second type is the optimization equation or the profit maximization equation of manufacturers. It describes how manufacturers get minimum cost or maximum profit under the constraint of production function, that is, remuneration of labor elements is equal to the marginal productivity. This type also determines manufacturers' demands for production factors.

2. Consumer Behavior

In the CGE model, consumers strive to achieve maximum utility under budget constraint. The consumer optimization problem is actually to choose the best combination of commodities (including service, investment, and entertainment) under budget constraint to achieve as high utility as possible. This aspect also has two types of equation: descriptive equation and optimization equation.

The descriptive equation introduces the budget constraint of consumers, that is, the disposable income of consumers. The optimization equation illustrates utility maximization behavior of consumers, which has various optional utility functions, such as Cobb–Douglas utility function, CES utility function, and the Stone–Geary utility function. Linear expenditure system (LES) or expanded linear expenditure system (ELES) equations that represent consumer expenses could be gained through derivation of utility function under budget constraint.

3. Government Behavior

There are many different methods in the CGE model that could be used to describe government behavior. Generally, the primary role of government is to make related policies, including policies concerning tax revenue, rate of interest, exchange rate, tariff, and fiscal subsidies. The CGE model introduces these policies into the equation system as exogenous control variables, so that it could affect different governmental macroeconomic policies in the whole economic system. Also, government is a consumer in the CGE model. Government revenues include tax and charges, and its expenses cover public utilities, transfer payment, fiscal subsidies, and government consumption.

4. Foreign Trade

The world economy is an open one, and the development of one country is closely related with that of other countries. Foreign trade has an important role in the economic development of one country. Therefore, foreign trade occupies an important position in the CGE model. One country with small import and output volume will not cause fluctuation of the world market price. The CGE model often assumes it as a small country and sets the world price of commodities as a constant. Additionally, it is necessary to distinguish important commodities from domestic ones. At present, the common way is to assume that domestic products and foreign products are imperfectly substituted. Import behaviors are described by the Armington hypothesis and CES equation; export behaviors are described by the constant elasticity transformation (CET) equation. Some departments can be viewed as having no import and export, import and export products and domestic products being either irreplaceable or partially replaceable.

5. Market Equilibrium

Describing market equilibrium is the key of the CGE model. Generally speaking, market equilibrium in the CGE model and the corresponding budget constraints include the following.

① Product market equilibrium: requires that the total supply of departments is equal to total demand in term of both quantity and value. If there is an imbalance in one department, the supply–demand difference can be treated as inventory. The product market equilibrium involving the inventory variable described by the CGE model is a generalized equilibrium.

② Factor market equilibrium: mainly refers to labor market equilibrium, that is, the total labor supply is equal to the total demand. Labor forces could flow among different departments to realize the optimization goals of manufacturers and consumers. Different marginal profit ratios of different departments mean that excessive labor supply during one period will surely lead to unemployment. So, labor forces flow among different departments to ensure labor market equilibrium. The labor market equilibrium in the CGE model refers to the generalized equilibrium involving unemployment.

③ Capital market equilibrium: mainly refers to the total investment being equal to total deposit. If the investment scale is inconsistent with the deposit level, offering of bonds, attraction of foreign capital, or changes in governmental fiscal reserves will be used as makeup strategies to regain the balance. Meanwhile, according to the theory of macroeconomics, the balance between investment and savings is equivalent to product market equilibrium because inventory in the product market is treated as one part of total investment and the other part is viewed as a fixed-asset investment. Some of the total savings will be used to buy inventory. Therefore, the product market equilibrium is in accordance with capital market equilibrium.

④ Government budget equilibrium: if government expenditures are not equal to government revenues, financial deficit will be added into the government revenue as a variable. Then, the government budget imbalance could be expressed by a group of equilibrium equations. Therefore, government revenues and expenditures also have characteristics of generalized equilibrium.

⑤ Resident incomes and expenditures: resident incomes include labor remuneration, enterprise profit distribution, and net foreign remittance. All incoming except individual income tax will be consumed or saved up to meet the balance of payments of residents.

⑥ International payment of balance: export surplus is expressed as inflow of foreign capital in the CGE model and import surplus is outflow of domestic capital. If the net inflow of foreign capital is treated as a variable, the international payment of balance shall meet the equilibrium relationship.

⑦ Relative equilibrium of resident utility: when studying relative population equilibrium among many regions, resident utility can be used as a measurement index of equality. If resident utility among regional population components reaches a relative equilibrium after mobility cost is deducted, these regions can be viewed as equal with each other. For this reason, it is necessary to introduce an equilibrium of resident utility.

In a word, changing the trend of balance variable under the foregoing equilibrium (e.g., inventory, unemployment, surplus, and deficit) provides an important basis for studying actual disequilibrium. Although the general equilibrium theory requires all equilibrium at the same time, this is proved impossible as a general rule and only conditional equilibrium could be realized. The macro-closure theory of the CGE model will explain it.

6. Macro-closure Theory of the CGE Model

The equilibrium solution is the premise of the CGE model. But how to ensure the existence, stability, and uniqueness of the model solution? The macro-closure theory of the CGE model is to solve this problem. Here, the basic meaning of closure is determination and the assignment of exogenous variables needed to solve a model. Different selections of exogenous variables and different choices of model closure reflect different hypotheses of factor market and macroscopic behaviors. In actual use, the CGE model has to be closed because (1) the typical CGE model does not include a currency mechanism and resource reallocation takes effect through a relative price mechanism rather than the absolute price mechanism; and (2) the typical CGE model is a constrained equilibrium model and macro-closure reflects such constraint (Zhen and Fan 1999).

In general, the closure rule must meet the following conditions. First, the number of equations is equal to the number of endogenous variables, which is accomplished by choosing endogenous and exogenous variables directly from model variables or by increasing or cutting equations. Second, there must be one variable in all price variables that is used as an exogenous variable, which is to determine a standardized price. Third, not all variables in one equation are exogenous ones to prevent it from being isolated. Fourth, only endogenous and exogenous variables with certain economic significance can be chosen. The chosen closure rules under different economic hypotheses will lead to different simulation results.

The key of the macro-closure theory is to promise an equal number of equations and variables in the CGE model. At the same time, it must conform to principles of economics. Therefore, different macro-closure theories reflect that the CGE model has different theoretical bases of economics that correspond to different research problems.

7.2.4 Multi-region CGE Model

Because the multi-region economy is universal nowadays, multi-region harmonious development becomes the hotspot of economics and regional science (Kim and Kim 2002; Giesecke 2002; Stifel and Thorbecke 2003). The multi-region CGE is a powerful tool to analyze multi-region economic problems. The following text introduces the basic structure of a typical CGE model concerning multi-region social equity in China. It is generally same as the universal CGE model except for some main equations. These main equations are modified according to the specific research goal. A population-grouping equation, resident utility equation, group income equation, central and local government demand, and an income equation and utility equilibrium equation are added.

The model integrates administrative regions in China into eight regions. It integrates the sector of national economy into eight production departments and assumes that every production department has one competitive enterprise and that each enterprise produces one commodity or provides one service. Each production department produces commodities or services for domestic market or exports with intermediate input, labor force, and capital input composed of composite goods (imported and domestic products use the Armington hypothesis and are exchanged though CES) (Fig. 7.9).

During the production process, the production department is not the decision maker of price, but the price taker. Therefore, enterprises (departments) must make production decisions according to preset maximum profit or minimum cost under certain technological conditions. Under the possible production boundary constraint, decision making shall follow the rule of maximizing revenue to determine relative shares of domestic market and foreign market.

Suppose, given the same scale, the total output of departments is beyond the control of manufacturers. It is determined by equilibrium conditions. In other words, manufacturers have to make a decision on input and choose appropriate intermediate input and factor effective input to minimize the production cost under a fixed total equilibrium condition (Fig. 7.10).

The model assumes that there are two types of family groups with different features (income, consumption, and investment) in China: urban residents and rural residents. Their consumption demand is composed of composite goods and is determined by budget constraint and the principle of utility maximization

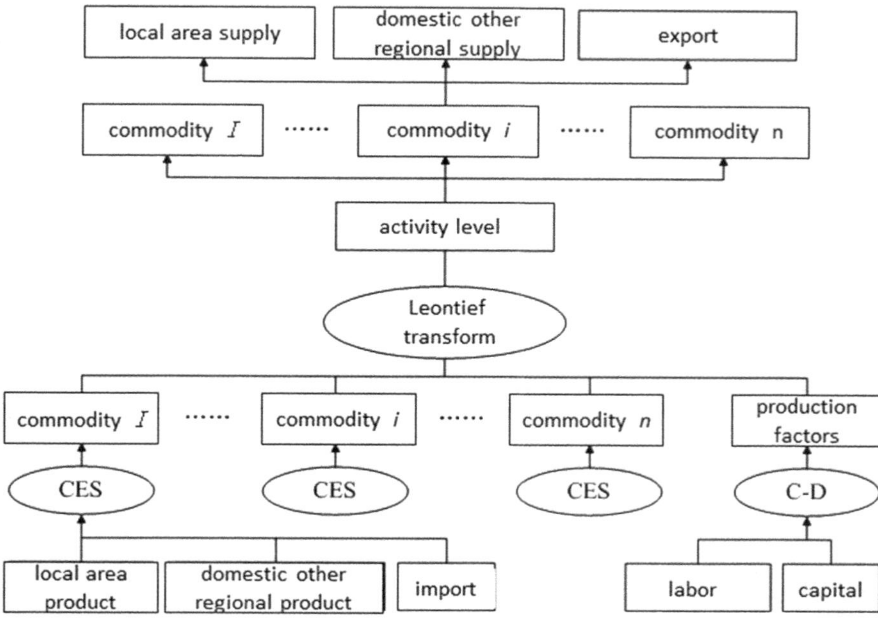

Fig. 7.10 Production structure of CGE model (Yanxin et al. 2008; Sun et al. 2009)

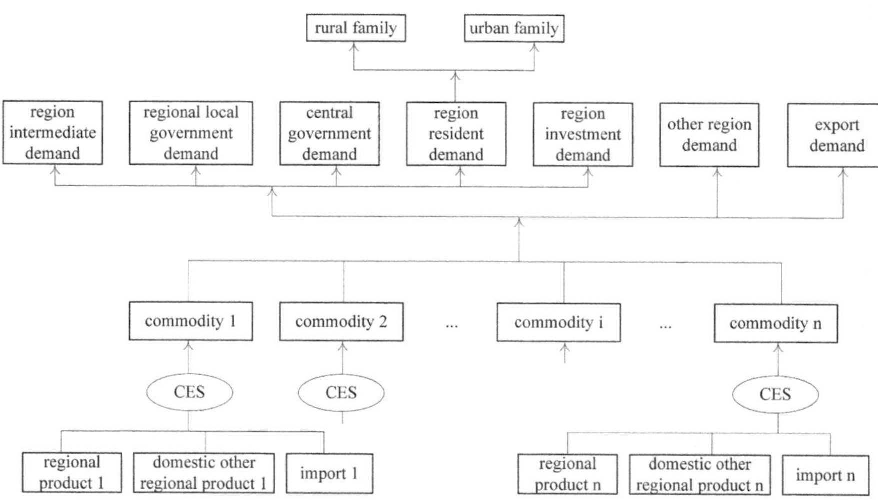

Fig. 7.11 Demand structure of the CGE model (Yanxin et al. 2008; Sun et al. 2009)

(Fig. 7.11). Common demand functions in the CGE model include the Linear Expenditure System (LES), Almost Ideal Demand System (AIDS), and an indirect utility function that surpasses the log function. All these demand functions could be derived by maximizing consumer utility of utility functions in certain forms.

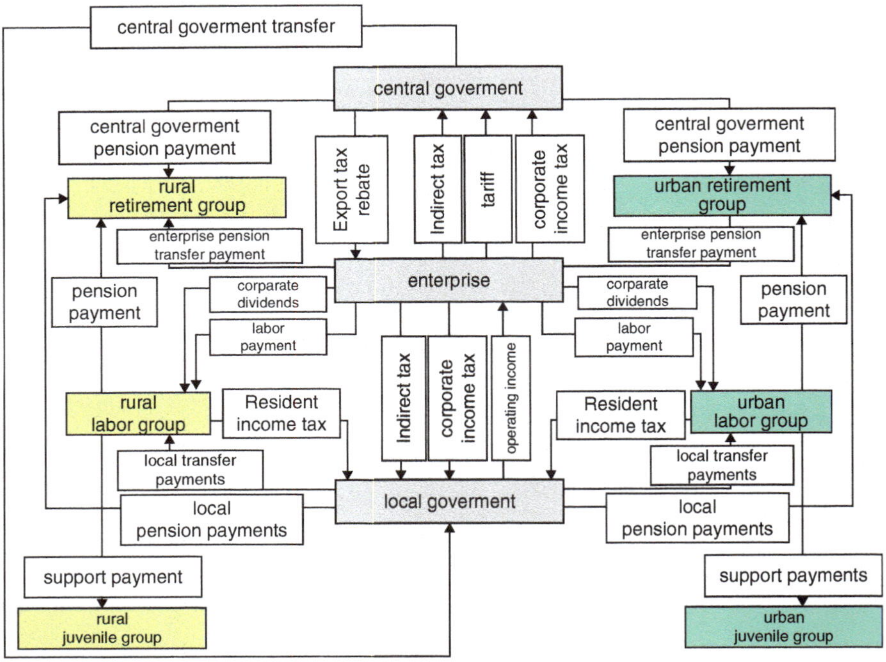

Fig. 7.12 Balance of payments of the CGE mode (Sun et al. 2009)

According to actual demand in China, the model closure adopts Kynesian mode in consideration of current huge unemployment and evident small demand in China. Labor supply can be viewed as infinite supply with fixed salary.

The CGE model not only divides central government and local governments, but also divides populations into a urban laborer group, urban retired group, urban young group, rural laborer group, rural retired group, and rural young group according to the urban and rural property and age structure. All are involved in income discussion (Fig. 7.12).

7.2.4.1 Multi-region CGE Model: Basic Equations

The multi-region CGE model of social equity mainly includes the product supply equation, composite good supply equation, labor and capital demand equation, import and export demand equation, government demand equation, resident demand equation, inventory change equation, resident income equation, enterprise revenue equation, government revenue equation, price system equation, savings equation, utility equation, and equilibrium condition equation.

The model divides equations into basic ones and expanded ones. Basic equations are introduced in the following text. Most of them exist in the general CGE model.

These basic equations in this multi-region CGE model of social equity are expanded to meet the needs of multi-region research.

Product Supply Equation

Differing from the single-region CGE model, the multi-region CGE model of social equality divides output further into value added, output of department *j* in region *i*, output sum of departments, and output sum of regions.

In the value-added equation, we use the Cobb-Douglas (C-D) production function. The C-D function has been widely used and has proved a good function form in practice. Also, its parameters are easy to estimate. The hypothesis of fixed scale is still used when determining the model.

1. Value added

$$VA_{ij} = A_{ij} L_{ij}^{\alpha_{ij}} K_{ij}^{1-\alpha_{ij}} \tag{7.2.1}$$

where VA_{ij} is value added in output of department j^* in region *i*; A_{ij} is all factor productivity; L_{ij} and K_{ij} are labor and capital inputs of department *j* in region *i*; and α_{ij} is labor elasticity substitution of department *j* in region *i*. This equation reflects that the model employs a hypothesis of constant scale and salary.

2. Output of departments in one region

$$X_{ij} = VA_{ij} + AIT_{ij} \tag{7.2.2}$$

where X_{ij} is output of department *j* in region *i*, which consists of VA_{ij} and intermediate input AIT_{ij}.

3. Output of departments in the whole country

$$ASX_j = \sum_{i=1}^{n} X_{ij} \tag{7.2.3}$$

where ASX_j is output of department *j* in the whole country, which is the sum of X_{ij} of all regions.

4. Total output of one region

$$ARX_i = \sum_{j=1}^{n} X_{ij} \tag{7.2.4}$$

where ARX_i is the total output of region *i*, which is the sum of outputs of all *j* departments in region *i*.

Composite Good Demand Equation
In this model, composite goods in one region include local products, products of other regions, and foreign products. The model applies the Armington hypothesis: domestic products and foreign products are different and have a certain imperfect substitutability, but domestic products have no difference and perfect substitutability. Therefore, products of the same department in the same region are priced equally.

For maximum utility, consumers could choose domestic products or imported products. The Armington hypothesis could be described by CES:

$$Q_{ij} = \xi_{ij}\left[\delta_{ij}M_{ij}^{-\rho_{ij}} + \left(1 - \delta_{ij}\right)D_{ij}^{-\rho_{ij}}\right]^{-\frac{1}{\rho_{ij}}} \tag{7.2.5}$$

where Q_{ij} is domestic demands of composite good j in region i; M_{ij} is import volume of commodity j in region i; D_{ij} is demands of domestic product j in region i; ξ_{ij} is a scale parameter; δ_{ij} is trade shares; and ρ_{ij} is elasticity of the substitution of trade.

Import Demand Equation
As in the single-region economic system, the multi-region economic system also has to consider the existence of international trade. The import demand equation can be gained from the first-order condition of consumer utility maximization (Xue 2006):

$$DM_{ij} = \left(\frac{\delta_{ij}}{1 - \delta_{ij}} \cdot \frac{PD_{ij}}{PM_{ij}}\right)^{\sigma_{ij}} \tag{7.2.6}$$

$$M_{ij} = DM_{ij} \cdot AD_{ij} \tag{7.2.7}$$

Equation (7.2.7) is the import demand equation. In Eq. (7.2.6), DM_{ij} is import/domestic product ratio; AD_{ij} is demands of domestic product j; M_{ij} is imports of product j; $\sigma_{ij} = \frac{1}{1+\rho}$ is elasticity of substitution between imports and exports; PD_j is price of domestic products; and PM_j is the price of imported products.

Export Demand Equation
In export demand function, we use the "small country hypothesis" according to the treatment of most CGE models of exports of developing countries. Hence, the export demands of departments in one region in this CGE model are assumed as the function of world market price/export price of the region commodity:

$$E_{ij} = \omega_{ij}\left(\frac{\overline{PWE_j}}{PE_j}\right)^{\eta_{ij}} \tag{7.2.8}$$

where E_{ij} is export volume of departments in one region; ω_{ij} is scale parameter; η_{ij} is elasticity of export price; $\overline{PWE_j}$ is world market price; and PE_j is export price.

Government Demand Equation

This CGE model distinguishes local government and central government, so government demands are also divided into central government demand and local government demand. It is worth pointing out particularly that local government in the model is a virtual regional government rather than the government of national administrative regions.

1. Local government demand:

$$\begin{aligned} RCG_{ij} &= \beta_{ij}^{rg} \cdot RCG_i^{tot}/P_{ij}, i = 1 \cdots 8 \\ \textstyle\sum_j \beta_{ij}^{rg} &= 1 \end{aligned} \tag{7.2.9}$$

2. Central government demand:

$$\begin{aligned} CCG_{ij} &= \beta_{ij}^{cg} \cdot CCG^{tot}/P_{ij} \\ \textstyle\sum_{i,j} \beta_{ij}^{cg} &= 1 \end{aligned} \tag{7.2.10}$$

Equations (7.2.9) and (7.2.10) represent local government demand and central government demand, respectively. Limited by data availability, we assume in the model system that local government only consumes local products whereas the central government could consume products of all regions. RCG_{ij} and CCG_{ij} are demands of product j of the local government and central government, respectively; RCG_i^{tot} and CCG^{tot} are total consumption of local government and central government; β_{ij}^{rg} and β_{ij}^{cg} are product consumption shares of local government and central government; and P_{ij} is product price. Government demand function can be gained by optimizing the government utility function (Wu and Wang 2004).

Inventory Change Equation

Inventory change is expressed by the proportion of inventory in total output:

$$ST_{ij} = kc_{ij} \cdot X_{ij} \tag{7.2.11}$$

Inventory change is presented in proportional distribution, kc_{ij} is proportionality factor, and ST_{ij} is inventory growth. Equation (7.2.11) reflects that inventory occupies a specific proportion in total output.

7.2.4.2 The Seven-Multi-region CGE Model: Specific Equations

In response to the existence of multiple economies, the multi-region CGE model has to expand specific equations according to the research goal (Sun et al. 2009).

Labor Demand Equation

The multi-region CGE model of social equality supposes that labor supply is adequate, which means that there is a certain underemployment rate in economy. Under the fixed wage rate, the labor supply is infinite and the total labor supply is fixed in the model. Common multi-region CGE models suppose that the national total labor supply is the sum of urban labor supply $L^{(u)*}$ and rural labor supply $L^{(r)}$ of the regions. However, this CGE model is to study social equality under multi-region conditions. Therefore, it is necessary to divide residents into several groups for adapting the target. In this model, first according to urban and rural property, residents are divided into urban residents and rural residents. Then, they are divided into five groups respectively according to ages: young group (0–15 years), laborer group 1 (16–32 years), laborer group 2 (33–45 years), laborer group 3 (46–64 years), and the retired group (≥65 years).[1]

$$P_i^{(u)} + P_i^{(r)} = Population_i \tag{7.2.12}$$

Equation (7.2.12) reflects that $Population_i$ of region i consists of urban residents $P_i^{(u)}$ and rural residents $P_i^{(r)}$.

$$\begin{aligned} P_i^{(u)} + P_i^{(r)} = {} & L_i^{(u1)} + L_i^{(u2)} + L_i^{(u3)} + R_i^{(u4)} + Y_i^{(u5)} + L_i^{(r1)} + L_i^{(r2)} \\ & + L_i^{(r3)} + R_i^{(r4)} + Y_i^{(u5)} \end{aligned} \tag{7.2.13}$$

$P_i^{(u)}$ is further divided into five age groups, including the young group $Y_i^{(u5)}$, urban laborer group 1 $L_i^{(u1)}$, urban laborer group 2 $L_i^{(u2)}$, urban laborer group 3 $L_i^{(u3)}$ and urban retired group $R_i^{(u4)}$. $P_i^{(r)}$ also is divided into five age groups, including the young group $Y_i^{(r5)}$, urban laborer group 1 $L_i^{(r1)}$, urban laborer group 2 $L_i^{(r2)}$, urban laborer group 3 $L_i^{(r3)}$and urban retired group $R_i^{(r4)}$.

$$L_i = \sum_{j=1}^{8} L_{ij} \tag{7.2.14}$$

Equation (7.2.14) reflects that labor force of region i is the sum of labor forces of all departments.

$$L_i^{(u)} + L_i^{(r)} = L_i \tag{7.2.15}$$

Equation (7.2.15) represents that labor force in region i is divided into $L_i^{(r)}$ and $L_i^{(u)}$ according to urban and rural property.

[1]In the following text, superscript h represent urban & rural attribute. The superscript u is urban and superscript r is rural unless there's specific notes.

$$L_i^{(u)} = L_i^{(u1)} + L_i^{(u2)} + L_i^{(u3)} \tag{7.2.16}$$

Equation (7.2.16) represents that $L_i^{(u)}$ is further divided into $L_i^{(u1)}$, $L_i^{(u2)}$, and $L_i^{(u3)}$ according to ages.

$$L_i^{(r)} = L_i^{(r1)} + L_i^{(r2)} + L_i^{(r3)} \tag{7.2.17}$$

Similarly, Eq. (7.2.17) shows that $L_i^{(r)}$ is divided into $L_i^{(r1)}$, $L_i^{(r2)}$, and $L_i^{(r3)}$.

Capital Demand Equation

Profit of enterprise is the gross trading income minus capital element returns, labor returns, and employee pensions:

$$\pi_{ij} = PN_{ij}X_{ij} - w_{ij}L_{ij} - r_{ij}K_{ij} - \delta_{ij}w_{ij}L_{ij} \tag{7.2.18}$$

where π_{ij} is profit of department j in region i; PN_{ij} is net price of product j in region i; w_{ij} is wage rate of department j in region i; r_{ij} is rate of capital return rate of department j in region i; and δ_{ij} is pension rate of department j in region i.

The enterprise goal is to maximize π_{ij}. According to the first-order condition of optimization:

$$PN_{ij} \cdot \alpha_{ij}A_{ij}L_{ij}^{\alpha_{ij}-1}K_{ij}^{1-\alpha_{ij}}/(1+\delta) = w_{ij} \tag{7.2.19}$$

$$PN_{ij} \cdot \left(1-\alpha_{ij}\right)A_{ij}L_{ij}^{\alpha_{ij}}K_{ij}^{-\alpha_{ij}} = r_{ij} \tag{7.2.20}$$

Then, the marginal value added of labor forced needed by enterprises is equal to the wage rate of labor force:

$$PN_{ij} \cdot \frac{\partial VA_{ij}}{\partial L_{ij}} = w_{ij} \tag{7.2.21}$$

Similarly, the marginal value added of capital shall be equal to the rental rate of capital:

$$PN_{ij} \cdot \frac{\partial VA_{ij}}{\partial K_{ij}} = r_{ij} \tag{7.2.22}$$

It can be concluded from Eqs. (7.2.21) and (7.2.22):

$$\frac{\partial VA_{ij}}{\partial L_{ij}} = \alpha_{ij}A_{ij}L_{ij}^{\alpha_{ij}-1}K_{ij}^{1-\alpha_{ij}}/(1+\delta) \tag{7.2.23}$$

$$\frac{\partial VA_{ij}}{\partial K_{ij}} = (1-\alpha_{ij})A_{ij}L_{ij}^{\alpha_{ij}}K_{ij}^{-\alpha_{ij}} \tag{7.2.24}$$

Resident Demand Equation

The resident demand equation uses the expanded linear expenditure system (ELES), a widely used demand function model.

$$\begin{gathered} P_{ij}\cdot ACOL_{ij}^{(h)} = P_{ij}\cdot\gamma_{ij}^{(h)} + \beta_{ij}^{(h)}\left(DIOL_i^{(h)} - \sum_j^n P_{ij}\cdot\gamma_{ij}^{(h)}\right) \\ 0<\beta_{ij}^{(h)}<1,\quad 0<\sum_i \beta_{ij}^{(h)} \le 1 \end{gathered} \tag{7.2.25}$$

$$\begin{gathered} P_{ij}\cdot ACOR_{ij}^{(h)} = P_{ij}\cdot\gamma_{ij}^{(h)} + \beta_{ij}^{(h)}\left(DIOR_i^{(h)} - \sum_j^n P_{ij}\cdot\gamma_{ij}^{(h)}\right) \\ 0<\beta_{ij}^{(h)}<1,\quad 0<\sum_i \beta_{ij}^{(h)} \le 1 \end{gathered} \tag{7.2.26}$$

In Eq. (7.2.25), $ACOL_{ij}^{(h)}$ is demand of the labor group h in region i for product j; $\gamma_{ij}^{(h)}$ is basic demands of the labor group h in region i for product j; $\beta_{ij}^{(h)}$ is expenditure proportion for the labor group h in region i for product j after the basic demands have been satisfied, which is equal to marginal propensity to consume product j; $DIOL_i^{(h)}$ is disposable income of the labor group h in region i; and P_{ij} is product price.

Similarly, in Eq. (7.2.26), $ACOR_{ij}^{(h)}$ is demand of the retired group h in region i for product j; $\gamma_{ij}^{(h)}$ is basic demand of the retired group h in region i for product j; $\beta_{ij}^{(h)}$ is the expenditure proportion of the retired group h in region i for product j after the basic demands have been satisfied, which is equal to marginal propensity to consume product j; and $DIOR_i^{(h)}$ is disposable income of the retired group h in region i.

According to research demand, this CGE model improved the general form of the enterprise revenue equation, government revenue equation, price system equation, and savings equation, and adds the income equation of different resident groups, the resident utility equation, and the equilibrium condition equation. Income equations mainly include enterprise revenue equation, resident income equation, and government revenue equation.

Enterprise Revenue Equation

Enterprises in the CGE model are divided according to regions and departments. In other words, there are eight enterprises in region i and every enterprise produces one product. Enterprise production follows the rule of maximum profit. The income equation of enterprise j in region i is

$$YE_{ij} = \left(PN_{ij} \cdot VA_{ij}^{(S)} - w_{ij} \cdot L_{ij} - \delta_{ij} w_{ij} \cdot L_{ij}\right)\left(1 - td_{ij}\right) + YGL_i \cdot ugs_{ij} \quad (7.2.27)$$

Enterprise revenue YE_i is net income. It is equal to the net price PN_{ij} multiplied by department output $VA_{ij}^{(S)}$ and minus remuneration for laborers (average wage rate w_{ij}, multiply laborer number L_{ij}), pensions (pension rate δ_{ij}, multiply remuneration for laborers), and income tax, and finally adds transfer payment of local government in region i to department j. In Eq. (7.2.27), YGL_i is the total local government revenue of region i; ugs_{ij} is proportion of transfer payment of local government to enterprises; and td_{ij} is corporate income tax rate.

Resident Income Equation

In the resident income equation system, the model considers per capita income of urban laborer groups, per capita income of the urban retired group, per capita income of rural laborer groups, and per capita income of the rural retired group. These data are added together to calculate the per capita income of residents.

Incomes of laborer groups include wages and property income. Disposable income of residents is equal to total income minus individual income tax and plus governmental transfer payment.

1. Per capita income of urban laborer groups

Per capita income of urban laborer groups comes from wages and property income.

$$inol_i^{(u)} = wol_i^{(u)} + etol_i^{(u)} \quad (7.2.28)$$

$$wol_i^{(u)} = \sum_{j=2}^{8} \mu_j w_{ij} \text{其中} \sum_{j=2}^{8} \mu_j = 1 \quad (7.2.29)$$

$$etol_i^{(u)} = \sum_{j=2}^{8} \mu_j YE_{ij} ueu_{ij} / \left(L_i^{(u1)} + L_i^{(u2)} + L_i^{(u3)} + R_i^{(u4)}\right) \quad (7.2.30)$$

According to Eq. (7.2.30), the per capita income of urban laborer groups in region i $inol_i^{(u)}$ include wages $wol_i^{(u)}$ and property income $etol_i^{(u)}$. $wol_i^{(u)}$ is gained from weighted average of wage rate of department j in region i, and $etol_i^{(u)}$ is gained from weighted average of a fixed proportion (ueu_{ij}) of enterprise output YE_{ij} in region i. μ_j is weight.

$$diol_i^{(u)} = inol_i^{(u)}\left(1 - \tau_i^{(u)} - \kappa_i^{(u)}\right)\left(1 - th_i^{(u)}\right) + lgt_i^{(u)} \quad (7.2.31)$$

$$lgt_i^{(u)} = YGL_i \cdot ugu_i / \left(L_i^{(u1)} + L_i^{(u2)} + L_i^{(u3)} + R_i^{(u4)}\right) \quad (7.2.32)$$

According to Eq. (7.2.31), the per capita income of urban laborer groups $diol_i^{(u)}$ includes two sources: one is per capita after-tax income, deducting pensions for the retired group and child-raising fee, and the other is governmental transfer payment to urban residents $lgt_i^{(u)}$. $\tau_i^{(u)}$ is the pension rate of urban laborer groups; $\kappa_i^{(u)}$ is the children-raising fee of urban laborer groups; and $th_i^{(u)}$ is the income tax rate of urban residents. It can be seen from Eq. (7.2.32) that local government transfer payment to urban residents, $lgt_i^{(u)}$, is a fixed proportion (ugu_i) of local government revenue YGL_i.

2. Per capita income of the urban retired group

Per capita income of the urban retired group includes pension transfer from laborer groups, pension transfer from enterprises, and transfer payments from local government and central government.

$$inor_i^{(u)} = ltor_i^{(u)} + etor_i^{(u)} + lgt_i^{(u)} + cgt_i^{(u)} \tag{7.2.33}$$

$$ltor_i^{(u)} = inol_i^{(u)}\left(L_i^{(u1)} + L_i^{(u2)} + L_i^{(u3)}\right)\tau_i^{(u)}/R_i^{(u4)} \tag{7.2.34}$$

$$\begin{aligned} etor_i^{(u)} = & \sum_{j=2}^{8} \mu_j w_{ij}\left(L_i^{(u1)} + L_i^{(u2)} + L_i^{(u3)}\right)\delta_{ij}/R_i^{(u4)} \\ & + \sum_{j=2}^{8} \mu_j YE_{ij} ueu_{ij}/\left(L_i^{(u1)} + L_i^{(u2)} + L_i^{(u3)} + R_i^{(u4)}\right) \end{aligned} \tag{7.2.35}$$

$$cgt_i^{(u)} = YG \cdot uglr_i/\left(R_i^{(u4)} + R_i^{(r4)}\right) \tag{7.2.36}$$

According to Eq. (7.2.32), the per capita income of the urban retired group $inor_i^{(u)}$ includes pension transfer from laborer groups $ltor_i^{(u)}$, pension transfer from enterprises $etor_i^{(u)}$, and transfer payments from local government $lgt_i^{(u)}$ and central government $cgt_i^{(u)}$. $cgt_i^{(u)}$ is a fixed proportion, $uglr_i$, of central government revenue YG.

3. Per capita income of the urban young group

$$ltoy_i^{(u)} = inol_i^{(u)}\left(L_i^{(u1)}\kappa_i^{(u1)} + L_i^{(u2)}\kappa_i^{(u2)} + L_i^{(u3)}\kappa_i^{(u3)}\right)/Y_i^{(u5)} \tag{7.2.37}$$

According to Eq. (7.2.37), the per capita income of the urban young group, $ltoy_i^{(u)}$, is the transfer payment from urban laborer groups.

4. Per capita income of rural laborer groups

Per capita income of rural laborer groups comes from wages and property income.

$$inol_i^{(r)} = wol_i^{(r)} + etol_i^{(r)} \tag{7.2.38}$$

$$wol_i^{(r)} = w_{i1} \tag{7.2.39}$$

$$etol_i^{(r)} = YE_{i1} uer_{i1} / \left(L_i^{(r1)} + L_i^{(r2)} + L_i^{(r3)} + R_i^{(r4)} \right) \tag{7.2.40}$$

According to Eq. (7.2.38), the per capita income of rural laborer groups, $inol_i^{(r)}$, includes wages $wol_i^{(r)}$ and property income $etol_i^{(r)}$ of rural laborers. $wol_i^{(r)}$is the wage rate of the first department (w_{i1}) and $etol_i^{(r)}$ is a fixed proportion uer_{i1} of income of the agricultural department YE_{i1}.

$$diol_i^{(r)} = inol_i^{(r)} \left(1 - \tau_i^{(r)} - \kappa_i^{(r)} \right) \left(1 - th_i^{(r)} \right) + lgt_i^{(r)} \tag{7.2.41}$$

$$lgt_i^{(r)} = YGL_i \cdot ugr_i / \left(L_i^{(r1)} + L_i^{(r2)} + L_i^{(r3)} + R_i^{(r4)} \right) \tag{7.2.42}$$

According to Eq. (7.2.41), the disposable income of rural laborer groups $diol_i^{(r)}$ has two sources: one is per capita after-tax income minus pension for the retired group and children-raising fee, and the other one is transfer payment from local government to rural residents$lgt_i^{(r)}$. $\tau_i^{(r)}$ is pension rate of rural laborer groups; $\kappa_i^{(r)}$is children-raising fee of rural laborer groups; and $th_i^{(r)}$ is income tax rate of rural residents.

Based on Eqs. (7.2.41) and (7.2.42), $lgt_i^{(r)}$is a fixed proportion ugr_i of income of local government revenueYGL_i. $\tau_i^{(r)}$ is the pension rate of rural laborer groups, and $\kappa_i^{(rm)}$ is children-raising fee of rural laborer groups, which varies for different groups.

5. Per capita income of the rural retired group

Per capita income of the rural retired group includes pension transfer from laborer groups, pension transfer from enterprises, and transfer payments from the local government and central government.

$$inor_i^{(r)} = ltor_i^{(r)} + etor_i^{(r)} + lgt_i^{(r)} + cgt_i^{(r)} \tag{7.2.43}$$

$$ltor_i^{(r)} = inol_i^{(r)} \left(L_i^{(r1)} + L_i^{(r2)} + L_i^{(r3)} \right) \tau_i^{(r)} / R_i^{(r4)} \tag{7.2.44}$$

$$\begin{aligned} etor_i^{(r)} = {} & w_{i1} \left(L_i^{(r1)} + L_i^{(r2)} + L_i^{(r3)} \right) \delta_{i1} / R_i^{(r4)} \\ & + YE_{i1} uer_{i1} / \left(L_i^{(r1)} + L_i^{(r2)} + L_i^{(r3)} + R_i^{(r4)} \right) \end{aligned} \tag{7.2.45}$$

$$cgt_i^{(r)} = YG \cdot uglr_i / \left(R_i^{(u4)} + R_i^{(r4)} \right) \tag{7.2.46}$$

It can be seen from Eq. (7.2.43) that the per capita income of the rural retired group $inor_i^{(r)}$ includes pension transfer from laborer groups $ltor_i^{(r)}$, pension transfer from enterprises $etor_i^{(r)}$ and transfer payments from local government $lgt_i^{(r)}$ and central government $cgt_i^{(r)}$. $cgt_i^{(r)}$ is a fixed proportion, $uglr_i$, of the central government revenue YG. w_{i1} is wage rate of the agricultural department in region i, and δ_{i1} is pension rate of the agricultural department.

6. Per capita income of the rural young group

$$ltoy_i^{(r)} = inol_i^{(r)}\left(L_i^{(r1)}\kappa_i^{(r1)} + L_i^{(r2)}\kappa_i^{(r2)} + L_i^{(r3)}\kappa_i^{(r3)}\right)/Y_i^{(r5)} \tag{7.2.47}$$

According to Eq. (7.2.47), the per capita income of the rural young group $ltoy_i^{(r)}$ is the transfer payment from rural laborer groups.

7. Total resident income of region i

The total resident income of region i IN_i includes the total urban resident income, $IN_i^{(u)}$, and the total rural resident income, $IN_i^{(r)}$.

$$IN_i = IN_i^{(u)} + IN_i^{(r)} \tag{7.2.48}$$

$IN_i^{(u)}$ covers all incomes of four urban resident groups:

$$IN_i^{(u)} = inol_i^{(u)}\left(L_i^{(u1)} + L_i^{(u2)} + L_i^{(u3)}\right) + inor_i^{(u)}R_i^{(u4)} \tag{7.2.49}$$

$IN_i^{(r)}$ covers all incomes of four rural resident groups:

$$IN_i^{(r)} = inol_i^{(r)}\left(L_i^{(r1)} + L_i^{(r2)} + L_i^{(r3)}\right) + inor_i^{(r)}R_i^{(r4)} \tag{7.2.50}$$

8. Total disposable income of residents in region i

The total disposable income of residents in region i DI_i includes the total disposable income of urban residents $DI_i^{(u)}$ and the total disposable income of rural residents $DI_i^{(r)}$.

$$DI_i = DI_i^{(u)} + DI_i^{(r)} \tag{7.2.51}$$

The total disposable income of residents covers all disposable incomes of five resident groups:

$$DI_i^{(h)} = DIOL_i^{(h)} + DIOR_i^{(h)} + DIOY_i^{(h)} \tag{7.2.52}$$

$$DIOL_i^{(h)} = diol_i^{(h)}\left(L_i^{(h1)} + L_i^{(h2)} + L_i^{(h3)}\right) \tag{7.2.53}$$

$$DIOR_i^{(h)} = inor_i^{(h)} R_i^{(h4)} \tag{7.2.54}$$

$$DIOY_i^{(h)} = inoy_i^{(h)} Y_i^{(h4)} \tag{7.2.55}$$

Government Revenue Equation

In the multi-region CGE model system, central government and local government shall be distinguished. The central government revenue mainly includes tariff revenues, indirect tax revenues, resident income tax revenues, and corporate income tax revenues. The export rebates of the central government can be viewed as a negative income of the government. The central government provides transfer payment to the local government. Indirect tax revenues, resident income tax revenues, and corporate income tax revenues are shared by the central and local government in proportion. The central government revenues minus transfer payment to local government are its disposable income. The local government revenues minus transfer payment to local residents and enterprises are its disposable income.

1. Tariff revenues of the central government

$$TM_i = \sum_j \left(\overline{PWM_j} \cdot M_{ij} \cdot tm_{ij} \cdot ER\right) \tag{7.2.56}$$

Equation (7.2.56) shows that the tariff revenues of the central government from products in region i (TM_i) are the sum of tariffs of various imported products [the product of world market price of imported products ($\overline{PWM_i}$), import tariff rate (tm_i), and exchange rate (ER)].

2. Export rebates of the central government

$$TE_i = \sum_j \left(\overline{PWE_j} \cdot E_{ij} \cdot te_{ij} \cdot ER\right) \tag{7.2.57}$$

Equation (7.2.57) shows that export rebates of the central government to exported products in region i (TE_i) are the sum of rebates for exported products [the product of world market price of exported products ($\overline{PWE_j}$), export rebate rate (te_j), and exchange rate (ER)].$TE_i$is the total export rebates of the government, which could be viewed as a negative income of the government.

3. Indirect tax revenues of the central government

$$TXC_i = \sum_j \left(PD_{ij} \cdot t_{ij} \cdot VA_{ij}^{(S)}.(1 - \phi_{ij})\right) \tag{7.2.58}$$

4. Indirect tax revenues of the local government

$$TXL_i = \sum_j \left(PD_{ij} \cdot t_{ij} \cdot VA_{ij}^{(S)} . \phi_{ij} \right) \tag{7.2.59}$$

Equation (7.2.59) reflects that the indirect tariff revenues of the central government from region i (TXC_i) are the sum of products of total domestic output $VA_{ij}^{(S)}$, domestic product price, PD_{ij} and indirect tax rate t_{ij} of local departments. Indirect tax revenues are shared between the central and local governments. ϕ_{ij} is the proportion of indirect tax revenues to the local government.

5. Corporate income tax revenues of the central government

$$TDC_i = \sum_j \left(PN_{ij} \cdot VA_{ij}^{(S)} - w_{ij} \cdot L_{ij} \right) \cdot td_{ij} . (1 - \eta_i) \tag{7.2.60}$$

It is known from Eq. (7.2.60) that the corporate income tax revenues of the central government from region i (TDC_i) are the product of net incomes of departments and the corporate income tax rate. Similarly, corporate income tax revenues are shared by the central and local governments. η_i is the proportion to the local government.

6. Corporate income tax revenues of the local government

$$TDL_i = \sum_j \left(PN_{ij} \cdot VA_{ij}^{(S)} - w_{ij} \cdot L_{ij} \right) \cdot td_{ij} . \eta_i \tag{7.2.61}$$

7. Resident income tax revenues of the local government

$$TDLH_i = \sum_h \left(W_i{}^{(h)} + ET_i{}^{(h)} \right) \cdot th_{ij} \tag{7.2.62}$$

Equation (7.2.62) shows that the resident income tax revenues of the local government from region i ($TDLH_i$) are the product of total resident incomes and the resident income tax rate.

8. Total revenues of the central government.

$$YG = \sum_i (TM_i + TXC_i - TE_i + TDC_i) \tag{7.2.63}$$

According to Eq. (7.2.63), total revenues of the central government (YG) are the sum of import tariff revenues collected from regions (TM_i), indirect tax revenues (TXC_i), and corporate income tax revenues (TDC_i) minus export debates (TDC_i).

9. Disposable incomes of the central government

$$DIG = YG \cdot \left(1 - \sum_i ugl_i - \sum_i uglr_i\right) \tag{7.2.64}$$

In Eq. (7.2.64), disposable incomes of the central government DIG are the total revenues of central government minus transfer payments to regions. ugl_i is a controllable policy variable and $uglr_i$ is the regular transfer payment of the central government to the regional retired.

10. Revenues of the local government

$$YGL_i = TXL_i + TDL_i + YG \cdot ugl_i + TDLH_i \tag{7.2.65}$$

In Eq. (7.2.65), revenues of the local government of region i (DIG) are the sum of indirect tax revenues, corporate tax revenues, resident tax revenues, and transfer payment from the central government.

11. Disposable incomes of the local government

$$DIGL_i = YGL_i \cdot (1 - ugr_i - ugu_i - ugs_i) \tag{7.2.66}$$

In Eq. (7.2.66), disposable income of the local government in region i (YGL_i) is the total revenue of the local government minus transfer payments to local urban residents, rural residents, and enterprises. The proportion of transfer payments of local government to local residents and enterprises is a controllable macro-policy variable.

Price Equation

In the price system, we apply the "small country hypothesis" to import and export prices. In other words, China is the accepter of price, but not the decision maker of price. The international market prices of imports and exports are defined as exogenous variables in the CGE model.

$$PM_{ij} = \overline{PWM_j} \cdot \left(1 + tm_{ij}\right) \cdot ER \tag{7.2.67}$$

Equation (7.2.67) reflects that domestic price of imported product (PM_{ij}) is determined by the fixed world import price $\overline{PWM_j}$, market exchange rate ER, and import tariff rate tm_{ij} together. In the CGE model, we assume that $\overline{PWM_j}$ is exogenous.

$$PE_{ij} = \overline{PWE_j} \cdot \left(1 + te_{ij}\right) \cdot ER \tag{7.2.68}$$

Equation (7.2.68) shows that the domestic price of exported product (PE_{ij}) is determined by the fixed world export price $\overline{PWE_j}$, market exchange rate ER, and export rebate rate te_{ij} together. Similarly, $\overline{PWE_j}$ is exogenous and domestic price could not influence the world price.

$$PD_{ij} = \left[\sum_h \sum_k a_{j,h}^{i,k} \cdot P_{kh} + \left(w_{ij} \cdot L_{ij} + r_{ij} \cdot K_{ij}\right)/VA_{ij}^{(S)}\right] / \left(1 - t_{ij}\right) \quad (7.2.69)$$

In Eq. (7.2.69), PD_{ij} is domestic product price of department j in region i; P_{kh} is composite product price; w_{ij} is wage rate; r_{ij} is rental rate of capital; L_{ij} is labor force of departments; K_{ij} is capital stock of departments; $VA_{ij}^{(S)}$ is total output of departments; t_{ij} is indirect tax rate; and $a_{j,h}^{i,k}$ is the Leontief coefficient.

$$PN_{ij} = PD_{ij}\left(1 - t_{ij}\right) - \sum_h \sum_k a_{j,h}^{i,k} \cdot P_{kh} \quad (7.2.70)$$

Because the supernormal profit of all departments is zero, PD_{ij} is the sum of intermediate input payment, labor cost, capital cost, and indirect tax. In Eq. (7.2.70), PN_{ij} is net price of department, which is equal to PD_{ij} minus intermediate input payment and indirect tax.

$$P_{ij} = \frac{AD_{ij}}{Q_{ij}} \cdot PD_{ij} + \frac{M_{ij}}{Q_{ij}} \cdot PM_{ij} \quad (7.2.71)$$

In Eq. (7.2.71), Q_{ij} is quantity of composite goods; M_{ij} is quantity of imported goods; and AD_i is quantity of domestic products. Price of Q_{ij} (P_{ij}) can be expressed as the weighted average of PD_{ij} and PM_{ij} (price of M_{ij}).

$$\sum_{i=1}^{n} \sum_{j=1}^{n} P_{ij} \cdot \Omega_{ij} = \bar{P} \quad (7.2.72)$$

Equation (7.2.72) represents that the weighted average of composite product price is a constant $\bar{P}$ and Ω_{ij} is price weight. When solving the model, $\bar{P}$ is set to 1.

Savings Equation

Savings equations include resident savings (urban resident savings and rural resident savings), business savings, and government savings (central government savings and local government savings), foreign savings, and out-of-region savings.

1. Resident savings

Resident savings is equal to resident disposable incomes minus consumption:

$$USAV_i = DIOL_i^{(u)} - \sum_j P_{ij} \cdot ACOL_{ij}^{(u)} + DIOR_i^{(u)} - \sum_j P_{ij} \cdot ACOR_{ij}^{(u)} \quad (7.2.73)$$

$$RSAV_i = DIOL_i^{(r)} - \sum_j P_{ij} \cdot ACOL_{ij}^{(r)} + DIOR_i^{(r)} - \sum_j P_{ij} \cdot ACOR_{ij}^{(r)} \quad (7.2.74)$$

Equation (7.2.73) shows that urban resident savings in region i ($USAV_i$) is equal to the disposable incomes of local urban residents $DIOL_i^{(u)} + DIOR_i^{(u)}$ minus total consumption $\sum_j P_{ij} \cdot ACOL_{ij}^{(u)} + \sum_j P_{ij} \cdot ACOR_{ij}^{(u)}$.

In Eq. (7.2.74), rural resident savings in region i ($RSAV_i$) are equal to the disposable incomes of local rural residents $DIOL_i^{(r)} + DIOR_i^{(r)}$ minus total consumption $\sum_j P_{ij} \cdot ACOL_{ij}^{(r)} + \sum_j P_{ij} \cdot ACOR_{ij}^{(r)}$.

2. Business savings

Business savings can be gained from Eq. (7.2.75). Business savings in region i (SE_i) are equal to local corporate incomes (YE_i) minus property expenditures of enterprises (ET_i).

$$SE_i = \sum_{j=1}^{8} \left(YE_{ij} - ET_{ij}\right) \quad (7.2.75)$$

3. Government savings

Government savings is government disposable income minus total government consumption. It includes savings of both the central and local governments. The central government savings (SG_i) is DIG_i minus total consumption of departments of all regions (CCG_{ij}):

$$SG_i = DIG_i - \sum\nolimits_j CCG_{ij} \quad (7.2.76)$$

The local government savings of region i (SGL_i) is $DIGL_i$ minus total consumption of local departments (RCG_{ij}):

$$SGL_i = DIGL_i - \sum\nolimits_j RCG_{ij} \quad (7.2.77)$$

4. Foreign savings

$$NSAV_i = \sum_{i,j} PM_{ij} \cdot M_{ij} - \sum_{i,j} PE_{ij} \cdot E_{ij} \quad (7.2.78)$$

where $NSAV_i$ is foreign savings. $\sum_{i,j} PM_{ij} \cdot M_{ij}$is total imports of regions and $\sum_{i,j} PE_{ij}$ $\cdot E_{ij}$ is total exports of regions.

5. Out-of-region savings

In the multi-region CGE model, there are savings of other regions (called the out-of-region savings) except for foreign savings. The out-of-region savings is the balance between import and export of the regions:

$$LSAV_i = \sum_{k,k \neq i} \sum_j IND_{kj} \cdot P_{ij} - \sum_{k,k \neq i} \sum_j OUD_{kj} \cdot P_{ij} \tag{7.2.79}$$

In Eq. (7.2.79), the out-of-region savings ($LSAV_i$) is the balance between total imports from other regions $\left(\sum_{k,k \neq i} \sum_j IND_{kj} \cdot P_j \right)$ and total exports to other regions $\left(\sum_{k,k \neq i} \sum_j OUD_{kj} \cdot P_j \right)$.

Resident Utility Equation

In this CGE model, resident utility is expressed by resident consumption.

1. Current utility of urban laborer groups in region i

$$UL_i^{(u)} = \rho_i \left(\sum_j ACOR_{ij}^{(u)} \right) \tag{7.2.80}$$

2. Current utility of rural laborer groups in region i

$$UL_i^{(r)} = \rho_i \left(\sum_j ACOR_{ij}^{(r)} \right) \tag{7.2.81}$$

3. Current utility of the urban retired group in region i

$$UR_i^{(u)} = \rho_i \left(\sum_j ACOR_{ij}^{(u)} \right) \tag{7.2.82}$$

4. Current utility of the rural retired group in region i

$$UR_i^{(r)} = \rho_i \left(\sum_j ACOR_{ij}^{(r)} \right) \tag{7.2.83}$$

5. Resident lifelong utility in region i

$$V_i^{(h)} = \sum_{m-1} \lambda^{m-1} UL_i^{hm} + \lambda^3 UR_i^{(h)} \tag{7.2.84}$$

Resident lifelong utility consists of two parts: laborer utility ($UL_i^{(h)}$) and retired utility $UR_i^{(h)}$. ρ_i is utility proportionality coefficient and λ is utility discount rate.

7.2.4.3 Multi-region CGE Model: Equilibrium Equation

Equilibrium conditions of the multi-region CGE model include capital market clearing, product market clearing and resident lifelong utility equilibrium.

Capital Market Clearing

$$\sum_i \sum_j K_{ij} = \bar{K} \tag{7.2.85}$$

Equation (7.2.85) shows that the national capital supply is equal to capital demand. It can be seen from Eq. (7.2.81) that capital in this model could flow within the whole country.

Investment–Savings Equilibrium

$$\sum_j ST_{ij} + \bar{K} = RSAV_i + USAV_i + SE_i + SG_i + SGL_i + NSAV_i + LSAV_i \tag{7.2.86}$$

Equation (7.2.86) shows that sum of capital supply $\bar{K}$ and stock $\sum_j ST_{ij}$ in region i is the sum of rural residents savings $RSAV_i$, urban residents savings $RSAV_i$, enterprise savings SE_i, savings of the central government in region iSG_i, savings of local government SGL_i, out-of-region savings $LSAV_i$, and foreign savings $NSAV_i$.

Product Market Clearing

According to market clearing conditions, the demands of the composite commodity in the region i (Q_{ij}) is the sum of intermediate input, household consumption, central government consumption, local government consumption, inventory growth, and export.

$$\sum_i \sum_j Q_{ij} = \sum_i \sum_j \Big(IT_{ij} + ACOL_{ij}^{(u)} + ACOR_{ij}^{(u)} + ACOL_{ij}^{(r)} + ACOR_{ij}^{(r)} + CCG_{ij} + RCG_{ij} + E_{ij} + ST_{ij} \Big) \tag{7.2.87}$$

Regional Resident Utility Equilibrium

$$\sum_{i-1}^{5} wghtoe_i V_i^{(h)} = \eta \mu_m \sum_{i-6}^{8} wghtow_i V_i^{(h)} \quad (7.2.88)$$

Equation (7.2.88) is regional resident lifelong utility equilibrium. When considering utility equilibrium, this model divides eight regions further into developed ones (northeast China, Beijing–Tianjin, north coastal region, east coastal region, and south coastal region) and underdeveloped ones (middle China, northwest China, and southwest China). It is believed that equilibrium is achieved when developed regions and underdeveloped regions have equal utility. In Eq. (7.2.88), $wghtoe_i$ is weight of developed regions and $wghtow_i$ is weight of underdeveloped regions.

According to Groenewold's model (2003), regions with equal resident utility are equal to each other and the model realizes equilibrium. Nevertheless, because there is a large regional difference, although complete utility equality has significance in theory, it makes little sense in policy. Therefore, equilibrium conditions of this multi-region CGE model mainly adopt the considerations of Zhu and Wang (2006). The resident ulitity difference of regions represnts by the parameter μ_m Because of the insurmountable mobility cost, regions can be viewed equal when their resident utility difference reaches at a very low level. μ_m and the model realizes equilibrium accordingly. However, this model involves an additional exogenous variable η to adjust utility difference among regions.

References

Adelman, I., & Robinson, S. (1978). *Income distribution policy in developing countries: A case study of Korea [M]*. Stanford: World Bank, Stanford University Press.

Alperovich, G. A. (1993). An explanatory model of city-size distribution: Evidence from cross-country data. *Urban Studies, 30*, 1591–1601.

Arrow K. J. (1951). An extension of the basic theorems of classical welfare economics[J]. In *Second Berkeley symposium on mathematical statistics and probability* (pp. 507–532).

Arrow, K. J., & Debreu, G. (1954). Existence of an equilibrium for a competitive economy [J]. *Econometrica: Journal of the Econometric Society*, 265–290.

Auerbach, F. (1913). Das gesetz der bevolkerungskoncentration. *Petermanns Geographische Mitteilungen, 59*, 74–76.

Backman, M., & Puu, T. (1990). *Spatial structure*. Springer.

Bandara, J. S. (1991). Computable general equilibrium models for development policy analysis in LDCs[J]. *Journal of Economic Surveys, 5*(1), 3–69.

Barro, R. J., & Sala-i-Martin, X. (2003). *Economic growth [M]* (2nd ed.). Boston: Mit Press Books.

Bayoumi, T., Coe, D. T., & Helpman, E. (1999). R&D Spillovers and global growth. *Journal of International Economics, 47*, 399–428.

Beckmann, M. J., & McPherson, J. C. (1970). City size distribution in a central place hierarch: An alternative approach. *J. Regional Sci., 10*, 25–33.

Ben-David, D. (1998). Free trade, growth, and convergence. *Journal of Economic Growth, 3*(2), 143–170.

Benenson, I., Omer, I., & Hatna, E. (2002). Entity-based modelling of urban residential dynamics-the case of Yaffo Tel Aviv. *Environment and Planning B, 29*(4), 491–512.

Bergman, L. (1990). Energy and environmental constraints on growth: A CGE modeling approach [J]. *Journal of Public Economics, 12*(4), 671–691.

Bloom, D., & Sachs, J. (1998). Geography, demography and economic growth in Africa. *Brookings Papers on Economic Activity., 2*, 207–295.

Braudel, F. (1994). *The characteristics of France: Space and history*. Beijing: The Commercial Press.

Brestchger, L. (1999a). *Growth theory and sustainable development*. Cheltenham: Edward Elgar Publishing Limited Glensanda House.

Brestchger, L. (1999b). Knowledge diffusion and development of regions. *The Annals of Regional Science, 33*, 251–268.

Cai, Y. L. (1995). The new ideas of the sustainable development of man –land system research. *Chinese Journal of Applied Ecology, 15*(3), 329–333.

Z. Wang, *Principles of Regional Science*, New Frontiers in Regional Science: Asian Perspectives 15, DOI 10.1007/978-981-10-5367-2

Caniëls, M. C. (2000). *Knowledge spillovers and economic growth*. Cheltenham: Edward Elgar.
Caniëls, M. C., & Verspagen, B. (2001). Barriers to knowledge spillovers and regional convergence in an evolutionary model. *Evolutionary Economics, 11*(3), 307–329.
Capozza, D. R., & Van Order, R. (1989). Spatial competition with consistent conjecture. *Journal of Regional Science, 29*(1), 1–13.
Chen, C., et al. (1991). *Principles of regional economic geography*. Science and technology of China Press.
Cheng, W., Sachs, J., & Yang, X. (2000). An inframarginal analysis of the Ricardian model. *Review of International Economics., 8*, 208–220.
China Statistical Yearbook. (2000). China Statistical Yearbook. China Statistics Press: Beijing.
Cohen, S. B. (1991). Global geopolitical change in the post-cold war era. *Annals of the Association of American Geographers, 81*(4), 551–580.
Conley, T. G., & Ligon, E. (2002). Economic distance and cross-country spillovers [J]. *Journal of Economic Growth, 7*, 157–187.
Dai, X. (2008). *Agent-based modeling of sequential innovation diffusion on network*. Shanghai: East China Normal University.
de Freitas, C. (1979). Human climates of northern China. *Atmospheric Environment, 13*, 71–77.
Demko, G. J., & Wood, W. B. (1994). *Reordering the world: Geopolitical perspectives on the twenty-first century*. Boulder: Westview Press.
Demurger, S., Sachs, J. D., Woo, W. T., et al. (2001). *Geography, economic policy, and regional development in China* (NBER Woring Paper Series 8897, pp 1–59).
Dervis, K., de Melo, J., & Robinson, S. (1982). *General equilibrium models for development policy [M]* (p. xvii, 526). Cambridge, New York: Cambridge University Press.
Desmet, K. (2000). A perfect foresight model of regional development and skill specialization. *Regional Science and Urban Economics, 30*, 221–242.
Diamond, J., & Ordunio, D. (2005). *Guns, germs and steel: The fates of human societies*. W.W. Norton & Company: New York.
Dixit, A., & Stiglitz, J. (1977). Monopolistic competition and optimum product diversity. *American Economic Review., 67*, 297–308.
Douven, R., & Peeters, M. (1998). GDP-spillovers in multi-country models [J]. *Economic Modelling, 15(2)*, 163–195.
Drezner, T. (1994). Locating a single new facility among existing unequally attractive facilities. *Journal of Regional Science, 34*, 237–252.
Eaton, J., & Eckstein, Z. (1997). Cities and growth: Theory and evidence from France and Japan. *Regional Science and Urban Economics, 27*(4), 434–474.
Ekins, P., Simon, S., Deutsch, L., Folke, C., & de Groot, R. (2003). Aframework for the practical application of the concept of critical natural capital and strong sustainability. *Ecological Economics, 44*, 165–185.
Ettlinger, N., & Archer, J. C. (1987). City-size distributions and the world urban system in the twentieth century. *Environment & Planning A, 19*, 1161–1174.
Fleming, M. (1962). Domestic financial policies under fixed and under floating exchanges rates. *International Monetary Fund Staff Papers, 9*, 369–379.
Frankel, J. (1988). Ambiguous policy multipliers in theory and in empirical models. In R. Bryant et al. (Eds.), *Empirical macroeconomics for interdependent economics*. The Brookings Institution: Washington.
Friesz, T. L., Miller, T., & Tobin, R. L. (1988). Algorithms for spatially competitive network facility location. *Environment and Planning B, 15*, 191–203.
Fujita, M., & Hu, D. P. (2001). Regional disparity in China 1985-1994: The effects of globalization and liberalization. *The Annals of Regional Science, 35*, 3–37.
Fujita, M., Krugman, P., & Venables, A. J. (1999). *The spatial economy: Cities, regions and international trade*. London: MIT Press.
Fujita, M., Krugman, P., & Venables, A. J. (2001). *The spatial economy:Cities, regions, and international trade*. Cambridge, MA: MIT Press.
Funke, M., & Strulik, H. (1999). Regional growth in West Germany: Convergence or divergence? *Economic Modelling, 16*, 489–502.

Gallup, J. L., Sachs, J. D., & Mellinger, A. (1999). Geography and economic development. *International Regional Science Review, 22*, 179–232.

Ghosh, A. R., & Masson, P. R. (1994). *Economic cooperation in an uncertain world [M]*. Cambridge, MA: Blackwell Publishers.

Giesecke, J. (2002). Explaining regional economic performance: an historical application of a dynamic multi-regional CGE model[J]. *Regional Science, 81*(2), 247–278.

Gong, Y., et al. (2013). Chinese industry structure evolution driven by innovation. *Studies in Science of Science, 31*(8), 1252–1259.

Goodfriend, M., & McDermott, J. (1995). Early development. *The American Economic Review, 85*, 116–133.

Green, J., & Scotchmer, S. (1995). On the division of profit in sequential innovation. *RAND Journal of Economics, 26*, 20–33.

Griliches, Z. (1979). Issues in assessing the contribution of R&D to productivity growth. *The Bell Journal of Economics, 10*, 92–116.

Grossman, G.,. M., & Helpman, E. (1991). Trade, knowledge spillovers, and growth. *European Economic Review, 35*, 517–526.

Gu, G., & Wang, Z. (2014). Agent-based simulation on the evolution and development of regional spatial structure with technology and capital diffusion in China. *Acta Geographica Sinica, 69*(6), 808–822.

Hall, R., & Jones, C. (1999). Why do some countries produce so much more output per worker than others. *The Quarterly Journal of Economics, 114*, 83–115.

Harris, R., & Cox, D. (1984). *Trade, industrial policy and Canadian manufacturing*. Toronto: University of Toronto Press.

Haynes, K., & Fotheringham, A. S. (1984). *Gravity and spatial interaction models*. Beverly Hills etc: SAGE.

Helliwell, J. F., & Padmore, T. (1985). Empirical Studies of macroeconomic interdependence. *Handbook of international economics*. Amsterdam: North-Holland.

Hetelling, H. (1929). Stability in competition. *Economics, 39*, 41–57.

Hojman, D. A., & Szeidl, A. (2008). Core and periphery in networks. *Journal of Economic Theory, 139*(1), 139295–139309.

Huang, R., Wang, Z., Deng J.-X., & Wu, J. (2015). Regional carbon emissions system development and application. *Scientia Geographica Sinica, 35*(4), 427–432.

Isard, W. (1975). *An introduction to regional science*. Englewood Cliffs: Prentice-Hall.

Jeong, B. (2002). Mesurement of human capital input across countries: A method based on the laborer's income. *Journal of Development Economics, 67*, 333–349.

Jeong, B. (2002). Mesurement of human capital input across countries: A method based on the laborer's income. *Journal of Development Economics, 67*, 333–349.

Jiang, Y. H. (2003). *The simulation study on Chinese CO2 emission reduction effect in endogenous economic growth framework*. Master dissertation.East China Normal University.

Jin, K. (2012). Industrial structure and capital flows. *The American Economic Review, 102*(5), 2111–2146.

Johansen, L. (1960). A multi-sectoral study of economic growth. *The Review of Economics and Statistics, 44*(2), 232–233.

Jones, E. L. (1981). *The European miracle: Environment, economies, and geopolitics in the history of Europe and Asia*. Cambridge: Cambridge University Press.

Kanaroglou, P., Liaw, K. L., & Papageorgiou, Y. Y. (1986). Analysis of migratory systems 1. Theory. *Environment and Planning A, 18*, 913–928.

Keller, W. J. (1980). *A general equilibrium approach [M]* (2nd ed., pp. 36–56). Amsterdam: North-Holland.

Kim, S. (1995). Region, resource and economics geography: Sources of U.S. regional comparative advantage, 1880–1987. *Regional Science and Urban Economics, 29*, 213–232.

Kim, E., & Kim, K. (2002). Impacts of regional development strategies on growth and equity of Korea: A multiregional CGE model[J]. *The Annals of Regional Science, 36*(1), 165–180.

Kohlhase, J. E., & Ohta, H. (1990). Löschian spatial competition and the related labor market. *Environment and Planning A, 22*, 1223–1238.

Krugman, P. (1991a). History and industry location: The case of the manufacturing belt. *American Economic Review, 81*, 80–83.

Krugman, P. (1991b). Increasing return and economic geography. *The Journal of Political Economy, 99*, 483–498.

Krugman, P. (1993a). First nature, second nature and metropolitan location. *Journal of Regional Science, 1993*, 129–144.

Krugman, P. (1993b). On the number and location of cities. *European Economic Review, 37*, 293–293.

Krugman, P. (1991). Increasing returns and economic geography. *Journal of Political Economy, 99*(3), 482–499.

Krugman, P. R. (1995). What do we need to know about the international monetary system. In P. R. Kenen (Ed.), *Understanding interdependence* (Vol. 81, pp. 509–529). Princeton University Press.

Landes, D. (1969). *The unbound Prometheus*. Cambridge: Cambridge University Press.

Leamer, E. E., et al. (1999). Does natural resource abundance increase Latin American income inequality. *Journal of Development Economics, 59*, 3–54.

Leontief, W. (1951). *The structure of American economy 1919–1939* (2nd ed.). New York: Oxford University Press.

Li, C. F. (1959). *The geographical structure of South America*. Beijing: Science Press.

Li, S., & Wang, Z. (2009). Computable models on the temporal and spatial scale of an optimum tourism destination circle. *Acta Geographica Sinica, 64*(10), 1255–1266.

Liu, Q. C., & Wang, Z. (2009). Research on geographric elements of economic difference in China. *Geographical Research, 28*(2), 430–440.

Liu, X., & Wang, Z. (2013). On the R&D Hub city in regional structure system. *China Soft Science*, (1), 98–107.

Lorentz, A., & Maria, S. (2010). *Structural change and business cycles: An evolutionary approach*. Papers on Economics and Evolution, Max Planck Institute of Economics, Evolutionary Economics Group: 1–40.

Lorentz, A., & Savona, M. (2008). Evolutionary micro-dynamics and changes in the economic structure. *Journal of Evolutionary Economics, 18*(34), 389–412.

Lösch, A. (1954). *The economics of location*. New Haven: Yale University Press.

Lu, D. D. (1989). *Location and regional research methods*. Beijing: Science Press.

Lu, D. D. (1995). *Regional development and spatial structure*. Beijing: Science Press.

Lu, D. D. (2001). An analysis of spatial structure and optimal regional development. *Acta Geographica Sinica, 56*(2), 127–135.

Lu, D. D. (2003). New factors and new patterns of regional development in China. *Geographical Research, 22*(3), 261–271.

Lucas, R. E. (1988). On the mechanics of economic development. *Journal of Monetary Economics, 22*(1), 3–42.

Lv, Z. K., & Wang, Z. (2008). A revisit of the size distribution of citiesL the case of China. *Modern Urban Research*, (6), 81–87.

Lyons, B., et al. (2001). Industrial concentration and market integration in the European Union. *Economica, 68*, 1–26.

Malmberg, A., Sölvell, Ö., & Zander, I. (1996). Spatial clustering, local accumulation of knowledge and firm competitiveness. *Geografiska Annaler Series B. Human Geography, 78*(2), 85–97.

Mao, H. Y., & Yu, D. L. (2001). Regional carrying capacity in Bohai rim. *Acta Geographica Sinica, 66*(3), 363–371.

Masters, W. A., & McMillan, M. S. (2001). Climate and scale in the economic growth. *Journal of Economic Growth.*, 167–186.

Mayhew, S., & Penny, A. (1992). *The concise Oxford dictionary of geography*. Oxford: Oxford University Press.

McKibbin, W. J., & Sachs, J. D. (1991). *Global linkages:Macroeconomic interdependence and cooperation in the world economy*. Washington: The Brookings Institution Press.

Mills, E. S., & Becker, C. M. (1986). *Studies in Indian urban development*. Oxford: Oxford University Press.
Molle, W. (1997). *The economics of European integration* (3rd ed.). Aldershot: Ashgate Publishing Limited.
Mundell, R. (1963). Capital mobility and stabilization policy under fixed and flexible exchange rate. *Canadian Journal of Economics and political Science, 20*, 475–85.
Nakagome, M. (1986). The spatial labour market and spatial competition. *Regional Studies, 20*, 307–312.
National Bureau of Statistics. (1999). Beijing: China Statistics Press.
Okahe, A., & Suzuki, A. (1987). Stability of spatial competition for a large number of firms on a bounded two-dimensionsal space. *Environment and Planning*.
O'Kelly, M. E., & Bryan, D. (1998). Hub location with flow economies of scale. *Transportation Research. Part B, 32*(8), 605–616.
O'Kelly, M. E., Bryan, D., Shorin-kapov, D., & Shorin-kapov, J. (1996). Hub network design with single and multiple allocation: A computational study. *Location Science., 4*(3), 125–138.
Onlin, B. (1933). *Interregional and international trade*. Cambridge, MA: Harvard University Press.
Palander, T. (1935). *Beitrage zur standortstheotie*. Stockholm: Almqrist and Wiksell.
Parr, J. B. (1987). The Tinbergen analysis of an urban system and alternative approaches. *Environment & Planning A, 19*, 187–204.
Pearce, D. W., & TurnerR, K. (1990). *Economicsof natural resources and the environment* (pp. 51–53). Baltimore: Johns Hopkins University Press.
Pereira, A. M., & Shoven, J. B. (1988). Survey of dynamic computational general equilibrium models for tax policy evaluation[J]. *Journal of Policy Modeling, 10*(3), 401–436.
Peter, M. Y. (1997). Humandomination of earth's ecosystems. *Science, 277*, 494–499.
Pflüger, M. (2004). A simple, analytically solvable, Chamberlinian agglomeration model. *Regional Science and Urban Economics, 34*(5), 565–573.
Pizer, W. A. (1998). The optimal choice of climate change policy in the presence of uncertainty. *Resource and Energy Economics, 21*, 255–287.
Porter, M. E. (1998a). Competitive advantage, agglomeration economies and regional policy. *International Regional Science Review., 19*(112), 85–90.
Porter, M. E. (1998b). Cluster and the new economics of competition. *Harvard Business Review, 76*(6), 77–90.
Porter, M. E. (2002). *The competitive advantage of nations* (Li Ming Xuan & Qiu Ru Mei, Trans.). Huaxia Press.
Pyka, A. (1997). Informal networking. *Technovation, 17*(4), 207–220.
Robinson, S. (1991). Macroeconomics, financial variables, and computable general equilibrium models[J]. *World Development, 19*(11), 1509–1525.
Roland-Holst, D., Reinert, K., et al. (1992). *North American Journal of Economics and Finance, 10*(1), 5–38.
Romer, P. M. (1986). Increasing return and long-run growth. *Journal of Political Economy, 94*, 1002–1037.
Rosen, K. T., & Resnick, M. (1980). The size distribution of cities: An examination of the Pareto law and primacy. *Journal of Urban Economics, 8*, 165–186.
Round, J. I., & Pyatt, G. (1985). *Social accounting matrices: A basis for planning[M]*. World Bank.
Roy, J. R., & Thill, J. C. (2003). Spatial interaction modeling. *Papers in Regional Science, 83*(1), 339–361.
Sante, I., Garcia, A. M., Miranda, D., & Crecente, R. (2010). Cellular automata models for the simulation of real-world urban processes: A review and analysis. *Landscape and Urban Planning, 96*, 108–121.
Scarf, H. (1973). *The computation of economic equilibria[M]*. New Haven: Yale University Press.
Shan, L., & Wang, Z. (2009). Computable models on regional spillovers of tourism industry: A case study of Yangtze River Delta. *Tourism Tribune, 24*(7), 18–26.

Shouji, C., & Liming, Z. (1998). *Fractal image compression*. Shanghai: Shanghai science and Technology Education Press.
Shoven, J. B., & Whalley, J. (1972). A general equilibrium calculation of the effects of differential taxation of income from capital in the US[J]. *Journal of Public Economics, 1*(3/4), 281–321.
Shoven, J. B., & Whalley, J. (1974). On the computation of competitive equilibrium on international markets with tariffs. *Journal of International Economics, 4*(4), 341–354.
Shoven, J. B., & Whalley, J. (1992). *Applying general equilibrium [M]*. Cambridge: Cambridge University Press.
Smith, T. E. (1987). Poissoa gravity models of spatial flows. *Journal of Regional Science, 27*, 315–340.
Soo, K. T. (2005). Zipf's law for cities: A cross-country investigation. *Regional Science and Urban Economics, 35*(3), 239–263.
Stevens, D., & Dragicevic, S. (2007). A GIS-based irregular cellular automata model of land-use change. *Environment and Planning B, 34*, 708–724.
Stifel, D. C., & Thorbecke, E. (2003). A dual-dual CGE model of an archetype African economy: Trade reform, migration and poverty[J]. *Journal of Policy Modeling, 25*(3), 207–235.
Sun, Y. W. Z., Zhu, Y. X., & Li, B. (2009). Controlling of regional gap based on general equilibrium theory. *Chinese Journal of Management, 6*(5), 1163–1168.
Tabuchi, T. (1986). Existence and stability of city-size distribution in the gravity and logit models. *Environment and Planning A, 18*, 1375–1389.
Teng, L., & Wang, Z. (2010). *Regional spillover*. Beijing: Science Press.
Teng, L., Wang, Z., Pang, L., & Li, G. Q. (2006). The effect of informatization on regional economy in China. *Human Geography, 21*(1), 72–75.
Terrasi, M. (1999). Convergence and divergence across Italian region. *The Annals of Regional Science, 33*(3).
Thomas, C., & Shelling. (1956). An essay on bargaining. *The American Economic Review, 6*.
Tinbergen, J. (1961). The spatial dispersion of production: A hypothesis Schweizerische Zeitscrift fue. *Volk Swirtschaft and Statisfik*, 412–419.
Tinbergen, J. (1968). The hierarchy model of the size distribution of centres. *Papers of Regional Science Association, 20*, 65–68.
Torrens, P. M. (2004). Geosimulation approaches to traffic modeling. In P. Stopher, K. Button, K. Haynes, & D. Hensher (Eds.), *Transport geography and spatial systems*. London: Pergamon.
van de Kilink, H. A. (1998). The pork network as a stage in port development: The case of Rotterdam. *Environment and Planning, A, 30*, 143–160.
Von Thunen, J. H. (1826). De r isolierte staat in, Beziehung auf landwirtschaft und nationalo konomie, Hamburg.
Wald, A. (1951). On some systems of equations of mathematical economics [J]. Econometrica: *Journal of the Econometric Society*, 368–403.
Walras, L., & Jaffé, W. (1954). *Elements of pure economics; or, The theory of social wealth[M]. London: American Economic Association and the Royal Economic Society.*
Wang, L. K. (1983). *Brownian motion and Newtonian potential*. Beijing: Science Press.
Wang, E. Y. (1988a). Geographical environment and social development. In S. X. Li & L. Ge (Eds.), *Human·nature·society* (pp. 146–156). Beijing: Peking University Press.
Wang, Z. (1988b). A class of geographic diffusion with self enhancement. *Young Geologist, TEG-1*(1), 40–44.
Wang, Z. (1995a). Spatial behavior of regional incentives. *Chinese Joutnal of Management Science, 2*(2), 9–15.
Wang, Z. (1995b). On the modern meaning of the relationship between the human and the earth. *10*(2), 1:5.
Wang, Z. (1998). The development of Chinese history reveals the sustainable development of the law. In L. J. Zhi et al. (Eds.), *The environment, resources and sustainable development in China facing to 21 century* (pp. 349–362). The Hong Kong Institute of Asia-Pacific Studies.
Wang, Z. (2000). Spatial interaction: A statistical mechanism model. *Journal of Geographical Sciences, 2000*(3), 87–92.
Wang, Z., & Ding, J. (1994). *Introduction to theoretical geography*. Beijing: Science Press.

Wang, Z., & Ge, Z. P. (2002). The multiple equilibrium states of China's regional economic development and the omen of changes. *Social Sciences in China, 4*, 31–38.

Wang, Z., & Peng, Y. (2012). Space operation of rural primary and secondary school location. *Acta Geographica Sinica, 68*(10), 1411–1417.

Wang, Z., & Sun, Y. (2013). A simulation on regional coordination and industrial structure evolution oriented to the main functional areas of China. *Scientia Geographica Sinica, 33* (6), 641–648.

Wang, Z., & Wang, L. (1998). The analysis on the ethics principle of sustainable development. In X. J. Min (Ed.), *The research of the discipline of human geography in China*. Beijing: Science Press.

Wang, Y., & Wang, Z. (2000). An analysis of tourism location. *Acta Geographica Sinica, 55*(3), 345–353.

Wang, Z., & Zhou, Q. (1995). A sudden climate change in the first half of nineteenth century. *Progress in Natural Science, 5*(3), 323–329.

Wang, Z., et al. (1991). On modern geography: Its objects, contents, structure and methodology. *Geographical Research, 10*(3), 68–77.

Wang, Z., Deng, F., et al. (1986). The spatial movement of mathematical simulation. *Young Geologist, z*(4), 40.

Wang, Z., Zhang, P. Y., & Zhou, Q. B. (1996). The impacts of climate on the society of China during historical times. *Acta Geographica Sinica, 51*(4), 329–339.

Wang, Z., Deng, Y., Ge, Z. P., et al. (2002). *Theoretical economic geography*. Beijing: Science Press.

Wang, Z., Liu, H. Y., & Liu, L. (2003a). A analysis to GDP-spillovers in China's 3 areas. *Economic Science*, (1), 5–13.

Wang, Z., Wang, Y., et al. (2003b). Research of reconstitution of tourism location of Guizhou province in China. *Geographical Research, 22*(3), 313–323.

Wang, Z., Ma, C., et al. (2004). A study on temporal and regional process of knowledge spillover. *The Annals of Regional Science, 38*, 595–606.

Wang, Z., Mao, K. J., et al. (2005). An analysis for location factors that cause industrial agglomeration. *Acta Geographica Sinica, 60*(4), 567–576.

Wang, Z. et al. (2007). GDP spillover simulation of CO2 emission reduction between China and the United State. *Acta Ecologica sinica, 27*(9), 3718–3726.

Wang, Z., Teng, L., & Cai, D. (2008a). A simulation to economic growth of two-region under resources restriction. *Journal of Natural Resources, 23*(4), 581–585.

Wang, Z., Zhu, Y. X., Mao, K. J., et al. (2008b). Two theoretical models for industrial cluster. *Science Research Management, 29*(5), 49–62.

Wang, Z., Li, H. Q., Wu, J., et al. (2010a). Policy modeling on the GDP spillovers of carbon abatement polices between China and the United States. *Economic Modelling, 20*, 40–45.

Wang, Z., Liu, T., & Dai, X.-Y. (2010b). Effect of policy and entrepreneurship on innovation and growth. *Studies in Regional Science, 40*(1), 19–26.

Wang, Z., Wu, J., Zhu, Y. B., et al. (2010c). *The economic research of climate protection*. Beijing: Science Press.

Wang, Z., Xia, H. B., & Wu, J. (2010d). *General geography* (pp. 98–99). Beijing: Science Press.

Wang, Z., Xue, J. B., Zhu, Y. B., Wu, J., & Zhu, Y. X. (2010e). *The CGE teconology of economic development simulation analysis*. Beijing: Science Press.

Wang, Z., Zhang, S., & Wu, J. (2012a). A new RICEs model with the global emission reduction schemes. *Chinese Science Bulletin, 57*(33), 4373–4380.

Wang, Z., Liu, C., & Mao, K. (2012b). Industry cluster: Spatial density and optimal scale. *The Annals of Regional Science, 49*, 719–731.

Wang, Z., Yao, Z., & Gu, G. X. (2014a). Multi agent-based simulation on technology innovation-diffusion in China. *Papers in Regional Science, 93*(2), 385–408.

Wang, Z., Sun, Y., & Gu, G. X. (2014b). Hub-network structure: new pattern for regional development management. *Bulletin of Chinese Academy of Sciences, 29*(3), 376–382.

Wang, Z., Han, Y., et al. (2016). The global geopolitical and Geopolinomical structure from the evolution of geographic natures. *Acta Geographica Sinica, 71*(6), 940–955.

Weber, A. (1909). Uber den standort der industrien, in English, Theory of the location of industries. Chicago: Unversity of Chicago Press.

Wei, H. K. (1997). Chinese regional economic growth and convergence. *China Industrial Economy, 3*, 31–37.

Weidlich, W., & Haag, G. (1987). A dynamic phase transition model for spatial agglomeration processes. *Journal of Regional Science, 27*, 529–569.

Whalley, J. (1985). *Trade liberalization among major world trading areas [M]* (pp. 69–82). Cambridge, MA: The MIT Press.

White, R., & Engelen, G. (1997). Cellular automata as the basis of integrated dynamic regional modeling. *Environment and Planning B, 24*, 235–246.

Williams, H. C., Kim, K. S., et al. (1990). Location—Spatial interaction models. *Environment and Planning A, 22*(1079–1089), 1155–1168.

Wilson, A. G. (1967). A statistical theory of spatial discribution models. *Transportation Research, 1*, 253–269.

Wilson, A. G. (1974). *Urban and regional model s in geography and planning*. London: Wiley.

Wilson, A. G., & Birkin, M. (1987). Dynamics models of agricultural location in a spatial interaction framework. *Geographical Analysis, 19*(1), 31–56.

Wu, C. J. (1981). Man – earth area system–The core of geographical study. *Economic Geography, 11*(3), 1–5.

Wu, B., & Wang, Z. (2004). The calculation and research on human capital of China's provinces. *Science Research Management, 25*(4), 60–65.

Wu, J., Wang, Z., & Xiong, W. (2008). Analysis of the influence of informatization level of China's industrial development. *Industrial Technology & Economy*, (7), 113–118.

Wu, J., Mahamod, R., & Wang, Z. (2011). Agent-based simulation of the spatial evolution of the historical population in China. *Journal of Historical Geography, 37*(1), 1–10.

Xia, H. B., & Wang, Z. (2012). Evolution of land differentiation in Chinese mainland. *Geographical Research, 31*(12), 2123–2138.

Xiong, W., & Wang, Z. (2012). New geopolinomic frame and economic crisis. *World Regional Studies, 21*(4), 1–12.

Xiong, W., Yao, Z. X., Wang, Z., & Sun, Y. (2010). Location choice of high-tech industry hub in China. *Studies in Science of Science, 28*(9), 1338–1346.

Xue, J. (2006). *Development and application of China's macroeconomic policy simulation based on CGE and its application*, [D]. Graduate University of Chinese Academy of Sciences, PhD thesis.

Yamamoto, K. (2008). Location of industry, market size, and imperfect international capital mobility. *Regional Science and Urban Economics, 38*(5), 518–532.

Yang, W. Y. (1987). *Location theory principle*. Gansu: Renmin Press.

Yang, X. K. (1999). Industrial division and specialization. In T. Min & M. Y. Shi (Eds.), *Modern economics front panel* (pp. 17–41). Beijing: The Commercial Press.

Yanxin, Z., Zheng, W., & Junb, X. (2008). The sector comparison of labor consumption in the technological progress. *Science Research Management., 29*(4), 111–118.

Ye, D. N., et al. (2001). Symmetric distribution of cities in China. *Science in China Series D*, (7), 608–616.

Zipf, G. K. (1949). *Human behaviour and the principle of least effort*. Reading: Addison-Wesley.

Zhen, Y., & Fan, T. (1999). *Chinese CGE model and policy analysis, [M]*. Beijing: Social Sciences Literature Press.

Zhou, W., & Wang, Z. (2003). China's regional technology guidance. *24*(6):132-137.

Zhu, Y., & Wang, Z. (2006). Research on the impact of transfer payment based on general equilibrium theory [J]. *China Management Science, 14*(Special issue), 394–308.

Index

Z. Wang, *Principles of Regional Science*, New Frontiers in Regional Science: Asian Perspectives 15, DOI 10.1007/978-981-10-5367-2

Printed by Printforce, the Netherlands